SYNTHESIZED TRANSMISSION LINES

SYNTHESIZED TRANSMISSION LINES

DESIGN, CIRCUIT IMPLEMENTATION, AND PHASED ARRAY APPLICATIONS

Tzyh-Ghuang Ma
National Taiwan University of Science and Technology, Taiwan

Chao-Wei Wang
MediaTek Inc., Taiwan

Chi-Hui Lai
ASUSTek Computer Inc., Taiwan

Ying-Cheng Tseng
National Taiwan University, Taiwan

WILEY

IEEE PRESS

This edition first published 2017

Registered Office
John Wiley & Sons, Singapore Pte. Ltd., 1 Fusionopolis Walk, #07–01 Solaris South Tower, Singapore 138628.

For details of our global editorial offices, for customer services and for information about how to apply for permission to reuse the copyright material in this book please see our website at www.wiley.com.

Library of Congress Cataloging-in-Publication Data

Names: Ma, Tzyh-Ghuang, 1973– author. | Wang, Chao-Wei, 1982– author. | Lai, Chi-Hui, 1985– author. | Tseng, Ying-Cheng, 1987– author.
Title: Synthesized transmission lines : design, circuit implementation, and phased array applications / Tzyh-Ghuang Ma, Chao-Wei Wang, Chi-Hui Lai, Ying-Cheng Tseng.
Description: Singapore ; Hoboken, NJ : John Wiley & Sons, 2017. | Includes bibliographical references and index.
Identifiers: LCCN 2016024736 (print) | LCCN 2016035371 (ebook) | ISBN 9781118975725 (cloth : alk. paper) | ISBN 9781118975749 (Adobe PDF) | ISBN 9781118975756 (ePub)
Subjects: LCSH: Microwave transmission lines–Design and construction. | Phased array antennas–Design and construction.
Classification: LCC TK7876 .S95 2017 (print) | LCC TK7876 (ebook) | DDC 621.3841/35–dc 3
LC record available at https://lccn.loc.gov/2016024736

A catalogue record for this book is available from the British Library.

Set in 10/12pt Times by SPi Global, Pondicherry, India
Printed and bound in Singapore by Markono Print Media Pte Ltd

10 9 8 7 6 5 4 3 2 1

To our beloved families and motherland

Contents

Preface **xi**

1 Introduction to Synthesized Transmission Lines **1**
C. W. Wang and T. G. Ma
1.1 Introduction 1
1.2 Propagation Characteristics of a TEM Transmission Line 2
1.2.1 Wave Equations 2
1.2.2 Keys to Miniaturization 5
1.3 Analysis of Synthesized Transmission Lines 7
1.3.1 Bloch Theorem and Characterization of a Periodic Synthesized Transmission Line 7
1.3.2 Characterization of a Non-Periodic Synthesized Transmission Line 9
1.3.3 Extraction of Line Parameters from S-Parameters 10
1.4 Lumped and Quasi-Lumped Approaches 11
1.4.1 Lumped Networks 11
1.4.2 Shunt-Stub Loaded Lines 14
1.5 One-Dimensional Periodic Structures 16
1.5.1 Complementary-Conducting-Strip Lines 19
1.6 Photonic Bandgap Structures 20
1.7 Left-Handed Structures 21
References 24

2 Non-Periodic Synthesized Transmission Lines for Circuit Miniaturization **26**
C. W. Wang and T. G. Ma
2.1 Introduction 26
2.2 Non-Periodic Synthesized Microstrip Lines and Their Applications 27
2.2.1 Design Details and Propagation Characteristics 27
2.2.2 90° and 180° Hybrid Couplers 30
2.2.3 Application to Butler Matrix as Array Feeding Network 32

2.3 Non-Periodic Synthesized Coplanar Waveguides and Their Applications 34
2.3.1 Synthesis and Design 34
2.3.2 180° Hybrid Using Synthesized CPWs 37
2.3.3 Dual-Mode Ring Bandpass Filters 38
2.4 Non-Periodic Quasi-Lumped Synthesized Coupled Lines 42
2.4.1 Basics of Coupled Transmission Lines 42
2.4.2 Miniaturization of Coupled Lines and the Directional Couplers 44
2.4.3 Marchand Baluns Using Synthesized Coupled Lines 49
2.4.4 Lumped Directional Coupler and the Phase Shifter 53
2.5 Non-Periodic Synthesized Lines Using Vertical Inductors 55
References 60

3 Dual/Tri-Operational Mode Synthesized Transmission Lines: Design and Analysis **62**
C. H. Lai and T. G. Ma
3.1 Introduction 62
3.2 Equivalent Circuit Models and Analysis 63
3.2.1 Ladder-Type Approximation in the Passband 63
3.2.2 Half-Circuit Model at Resonance 64
3.3 Dual-Operational Mode Synthesized Transmission Lines 65
3.3.1 Design Concept 65
3.3.2 Dual-Mode Synthesized Line Using a Series Resonator 66
3.3.3 Dual-Mode Synthesized Line Using Open-Circuited Stubs 70
3.3.4 Dual-Mode Synthesized Line Using Parallel Resonators 72
3.4 Tri-Operational Mode Synthesized Lines Using Series Resonators 74
3.4.1 Design Concept 74
3.4.2 Tri-Mode Synthesized Line as Category-1 Design 75
3.4.3 Tri-Mode Synthesized Line as Category-2 Design 79
3.4.4 Tri-Mode Synthesized Line as Category-3 Design 83
3.5 Multi-Operational Mode Synthesized Lines as Diplexer and Triplexer 87
3.5.1 Diplexer 87
3.5.2 Triplexer 89
References 94

4 Applications to Heterogeneous Integrated Phased Arrays **95**
C. H. Lai and T. G. Ma
4.1 Introduction 95
4.2 Dual-Mode Retrodirective Array 96
4.2.1 Design Goal 96
4.2.2 System Architecture 97
4.2.3 Circuit Realization 98
4.2.4 Bistatic Radiation Patterns 102
4.2.5 Alternative Architecture 103
4.3 Dual-Mode Integrated Beam-Switching/Retrodirective Array 106
4.3.1 Design Goal 106
4.3.2 System Architecture 106

4.3.3 *Circuit Realization* 109
4.3.4 *Radiation Characteristics* 111
4.3.5 *Complementary Design* 111
4.4 Tri-Mode Heterogeneous Integrated Phased Array 115
4.4.1 *Design Goal* 115
4.4.2 *System Architecture* 116
4.4.3 *Operation and System Implementation* 117
4.4.4 *Circuit Responses and Radiation Patterns* 119
4.4.4.1 *Beam-Switching Mode* 120
4.4.4.2 *Van Atta Mode* 122
4.4.4.3 *PCA Mode* 122
4.5 Simplified Dual-Mode Integrated Array Using Two Elements 122
References 124

5 On-Chip Realization of Synthesized Transmission Lines Using IPD Processes 126
Y. C. Tseng and T. G. Ma
5.1 Introduction 126
5.2 Integrated Passive Device (IPD) Process 127
5.3 Tight Couplers Using Synthesized CPWs 128
5.3.1 *Quadrature Hybrid* 128
5.3.2 *Wideband Rat-Race Coupler* 129
5.3.3 *Dual-Band Rat-Race Coupler* 132
5.3.4 *Coupled-Line Coupler* 137
5.3.5 *Butler Matrix* 139
5.4 Bandpass/Bandstop Filters Using Synthesized CPWs 142
5.4.1 *Bandpass Filter Using Synthesized Stepped-Impedance Resonators* 143
5.4.2 *Transformer-Coupled Bandpass Filter* 146
5.4.3 *Bridged T-Coils as Common-Mode Filter* 147
5.5 Chip Designs Using Multi-Mode Synthesized CPWs 151
5.5.1 *Diplexer* 151
5.5.2 *Dual-Mode Rat-Race Coupler* 154
5.5.3 *Triplexer* 157
5.5.4 *On-Chip Liquid Detector* 161
References 166

6 Periodic Synthesized Transmission Lines with Two-Dimensional Routing 168
T. G. Ma
6.1 Introduction 168
6.2 Design of the Unit Cells 169
6.2.1 *Formulation* 169
6.2.2 *Quarter-Wavelength Lines* 172
6.3 Power Divider and Couplers 174
6.4 Broadside Directional Coupler 178
6.4.1 *Design Principle* 178
6.4.2 *Circuit Realization* 180

6.5 Common-Mode Rejection Filter 184
6.5.1 Design Principle 184
6.5.2 Circuit Realization 187
6.6 On-Chip Implementation 189
6.6.1 Unit Cells and Quarter-Wavelength Lines 189
6.6.2 Circuit Implementations and Compensation 192
References 194

Index **196**

Preface

This book intends to provide comprehensive coverage of the recent progress in synthesized (or artificial) transmission lines for graduate students in electrical and telecommunication engineering. Synthesized transmission lines are microwave lumped or quasi-lumped networks that have similar electrical properties to a uniform transmission line, but are of a far more compact size. This unique feature makes this sort of microwave wave-guiding structure an ideal candidate for realizing miniaturized microwave passive components with comparable performances to their conventional counterparts. Add-on values such as harmonic suppression, non-integer ratio between passbands, multi-functional operation, and so on, are demonstrated through the years.

The first part of this book focuses on introducing basic synthesis techniques and analysis tools for developing synthesized transmission lines with or without periodicity. Classical approaches are introduced along with simple examples for easy understanding. The basic principles are followed by a variety of synthesized transmission lines in microstrip, coplanar waveguide, or stripline form, and their applications to miniaturized passive components including couplers, array feeding networks, filters, and phase shifters.

The second part of this book is devoted to providing a comprehensive introduction to a new sort of wave-guiding structure, termed *multi-operational mode synthesized transmission lines*. This is the result of 10 years of research work conducted by the authors at the National Taiwan University of Science and Technology, Taiwan. Multi-operational mode synthesized transmission lines, abbreviated to multi-mode synthesized lines, can provide distinct electrical properties at different frequencies or in different material media. Without using active switches, the synthesized line could be identical to a uniform transmission line in one band, but auto-configures as an open or short circuit in another band. A variety of applications not feasible with conventional microwave components, including multiplexers and multi-mode feeding networks for phased arrays, are introduced.

The third part of the book provides thorough coverage of recent on-chip development of synthesized transmission lines using an emerging fabrication technology, the integrated passive device (IPD) process. The IPD process is a competitive technology in the integration of on-chip microwave passive and active components for system-in-package (SiP) applications. This book will be the first book dedicated to summarizing state-of-the-art on-chip components using synthesized transmission lines with IPD technology.

The final part of the book covers a new sort of one-dimensional periodic synthesized transmission line with two-dimensional routing capability. It is also an outcome of the research conducted by the author group. For the most part of this chapter, designs are disclosed to the public for the first time. The periodic synthesized transmission lines make the routing of a passive microwave component conformal to an arbitrary outline profile when integrated with other circuit modules in the same system.

In the course of preparing the book, a number of people offered their kind support and assistance. First of all, the authors would like to express their appreciation to Dr. Jenshan Lin at University of Florida for his kind support of the environment in which Tzyh-Ghuang Ma worked on the book. Also, the financial support from the Ministry of Science and Technology, Taiwan for covering his living expenses in Gainesville, Florida, is highly appreciated. The authors would also like to express their sincere gratitude to Dr. Tzong-Lin Wu at National Taiwan University, and Dr. Zhi Ning Chen at National University of Singapore for their continuous encouragement and inspiring thoughts. Without their great support, this book could not have been written.

The authors are indebted to many researchers for their excellent work published in IEEE journals; they enrich the scope of the book and the sources of reference. The authors are grateful to the editorial staff of John Wiley & Sons, Ltd for their great efforts in the preparation of the book, from project management to proofreading. In particular, they are grateful to James Murphy, Maggie Zhang, Nivedhitha Elavarasan, Shivana Raj, Lynette Woodward, Karthika Sridharan, Victoria Taylor, and Clarissa Lim. The authors are especially grateful to the staff at National Chip Implementation Center (CIC), Taiwan for fabricating and probing chips over the years.

In addition, the authors are grateful to the former and present team members, Chao-Wei Wang, Yu-Ting Cheng, Chen-Cheng Wang, Chi-Hui Lai, Hung-Wei Hsu, Jyh-Woei Tsai, Huang-Chin Chiu, Cheng-Hsun Wu, Kai-Chi Lin, Yin-Cheng Tseng, Jyun-Yu Zou, Wei-Chen Wang, Chen-Yuan Shiau, Wei-Shan Chung, Bo-Feng Su, Guan-Ting Zhou, Chang-An Lin, Yi-Ling Wu, Ling Chieh Hsu, Hung-Hsuan Chen, Li Chang, Huy Nam Chu, Ting Hsieh, Hua-Chien Liao, and Gao-Yi Li, at the Applied Electromagnetics Lab at National Taiwan University of Science and Technology for their consistent hard work over the years in exploring the field of synthesized transmission lines. It has been 10 years since the authors first became devoted to this special topic, but they still find new things out! They also wish to thank Jhen-Jia Liao, Gao-Yi Li, Hua-Chien Liao, Hong-Yi Yang, Yen-Yin Chen, and Kuan-Ting Fang for redrawing illustrations and proofreading, and You-Jiun Wang, Min-Chung Chen, Guan-Yi Li, Chia-Chun Kao and Taufiqqurrachman for proofreading.

Finally, the authors are deeply indebted to their family for their unconditional love and understanding. Tzyh-Ghuang Ma would also like to express his deepest appreciation to his wife, Ching-Wen, for her patience and great care during the writing of the book. Chao-Wei Wang would like to express his deep appreciation to his wife, Yiling, for her love and encouragement.

This book is dedicated to everyone who works hard over the years for our country. God bless Formosa.

Tzyh-Ghuang Ma
Taipei, Taiwan

1

Introduction to Synthesized Transmission Lines

C. W. Wang and T. G. Ma

1.1 Introduction

In modern communication systems, the rapid evolution of integrated circuit (IC) and packaging technologies have driven more and more function blocks to be integrated into a single chip/module. In the second decade of the twenty-first century, highly-integrated front-end modules such as microwave/millimeter-wave radar and image systems, [1–4], phased arrays [5, 6], and so on have hit the commercial markets. In general, the RF modules require a large number of transmission-line-based elements for vector signal processing in the analog domain. The transmission line elements, however, inevitably occupy a large circuit area. In the cost-driven market, area is the cost. It therefore leads to an enormous amount of research work focusing on developing various kinds of synthesized transmission lines for reducing the required circuit size. A synthesized transmission line is a lumped or quasi-lumped network that may function identically to a uniform transmission line within a given bandwidth.

Synthesized transmission lines can be developed with or without periodicity. In a broad sense, it could be either right-handed or left-handed depending on the forming blocks. To describe the general concept, in this chapter we will start from Maxwell's equations and discuss the analog between plane wave propagation in a material media and the TEM mode in a parallel-plate waveguide. The parameters associated with the wave propagation and their corresponding circuit parameters in a transmission line are linked herein. Based on the fundamental principle, design formulae for periodic and non-periodic synthesized transmission lines are summarized. Classical design approaches are reviewed to demonstrate how synthesized transmission lines are realized practically. A brief review of left-handed synthesized lines, or metamaterial structures, is provided at the end of the chapter.

Synthesized Transmission Lines: Design, Circuit Implementation, and Phased Array Applications, First Edition.
Tzyh-Ghuang Ma, Chao-Wei Wang, Chi-Hui Lai and Ying-Cheng Tseng.

The formulae in this chapter form the basis of the non-periodic synthesized transmission lines in Chapter 2 for circuit miniaturization, and in Chapter 5 for chip implementation. The multi-operational mode synthesized transmission lines in Chapters 3 and 4, for phased array applications, are also derived using the same building blocks. The two-dimensional synthesized transmission lines in Chapter 6 also follow the periodic condition in Sec. 1.3.1.

1.2 Propagation Characteristics of a TEM Transmission Line

In this section, we start with the Maxwell's equations to derive the governed equations in a wave-guiding structure under the assumption of a transverse electromagnetic (TEM) field distribution. The propagation characteristics are summarized and compared to a distributed transmission line having similar mathematical forms by using circuit parameters.

1.2.1 Wave Equations

As shown in Fig. 1.1, consider a parallel-plate waveguide operated in the TEM mode. The field distribution inside the wave-guiding structure is known to be identical to a uniform plane wave in free space with uniquely defined voltage and current in the transverse plane. Maxwell's curl equations in a source-free region are:

$$\nabla \times \vec{E} = -\mu \frac{\partial \vec{H}}{\partial t}, \tag{1.1}$$

$$\nabla \times \vec{H} = \varepsilon \frac{\partial \vec{E}}{\partial t}. \tag{1.2}$$

Assuming that the wave propagates along the z-direction, the fields transverse to the direction of propagation in a parallel-plate waveguide, from (1.1) and (1.2), are:

$$\frac{-\partial E_x}{\partial z} = \mu \frac{\partial H_y}{\partial t}, \tag{1.3}$$

$$\frac{-\partial H_y}{\partial z} = \varepsilon \frac{\partial E_x}{\partial t}. \tag{1.4}$$

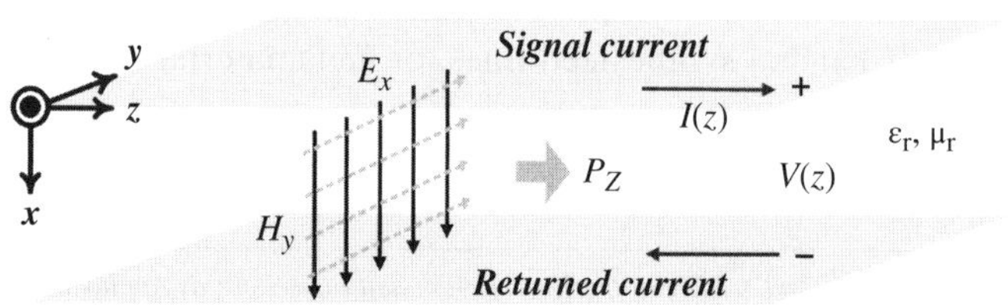

Figure 1.1 Wave propagation in a parallel-plate waveguide

Partially differentiate (1.3) with respect to z and (1.4) with respect to t to get,

$$\frac{-\partial^2 E_x}{\partial z^2} = \mu \frac{\partial^2 H_y}{\partial z \partial t}, \tag{1.5}$$

$$\frac{-\partial^2 H_y}{\partial t \partial z} - \varepsilon \frac{\partial^2 E_x}{\partial t^2}, \tag{1.6}$$

Substitution of (1.6) into (1.5) yields

$$\frac{\partial^2 E_x}{\partial^2 z} = \mu\varepsilon \frac{\partial^2 E_x}{\partial^2 t}. \tag{1.7}$$

It is a second-order partial differential equation known as the *one-dimensional wave equation*, which can be applied to any wave-guiding structure supporting TEM wave propagation. The phase velocity of the TEM wave is simply

$$v_p = \frac{1}{\sqrt{\mu\varepsilon}}. \tag{1.8}$$

where μ and ε are the permeability and permittivity of the medium filled within the wave-guiding structure.

Now, let us turn our attention to a lossless distributed uniform transmission line modeled by periodically loaded *LC* sections, as shown in Fig. 1.2. Under the assumption that each lumped *LC* segment is infinitesimal in length, the voltage and current along the line, from Kirchhoff's laws, are related to each other by,

$$\frac{-\partial V}{\partial z} = L \frac{\partial I}{\partial t}, \tag{1.9}$$

$$\frac{-\partial I}{\partial z} = C \frac{\partial V}{\partial t}. \tag{1.10}$$

L and C are the per-unit-length inductance and capacitance of the line. Equations (1.9) and (1.10) are known as the *Telegrapher's equations* and actually take the same form as (1.3) and (1.4).

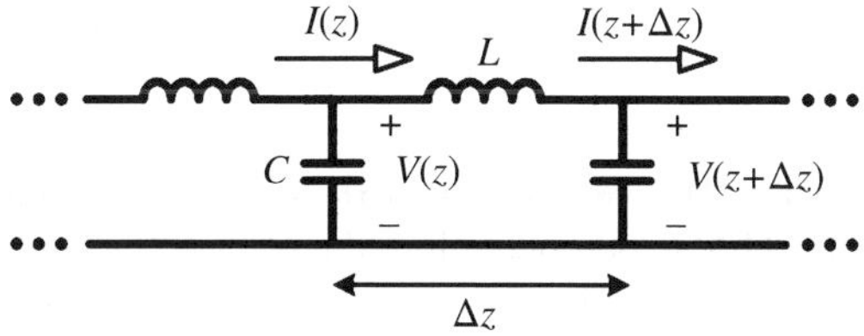

Figure 1.2 Equivalent lumped *LC* model of a distributed uniform transmission line

Following the same mathematical procedure, the wave equation is derived in terms of the voltage (*V*) or current (*I*) as,

$$\frac{\partial^2 V}{\partial^2 z} = LC\frac{\partial^2 V}{\partial^2 t}. \tag{1.11}$$

The general solution of the voltage and current waves propagated along the lossless uniform transmission line, from (1.11), is

$$V(z) = V^+e^{-j\beta z} + V^-e^{+j\beta z}, \tag{1.12}$$

$$I(z) = I^+e^{-j\beta z} - I^-e^{+j\beta z}. \tag{1.13}$$

β is known as the *phase constant* or *guided wavenumber*,

$$\beta = \frac{\omega}{v_p} = \omega\sqrt{LC}. \tag{1.14}$$

The phase velocity of the voltage and current waves is therefore,

$$v_p = \frac{1}{\sqrt{LC}}. \tag{1.15}$$

Meanwhile, differentiating (1.12) with respect to z, we have

$$\frac{\partial V(z)}{\partial z} = -j\beta V^+e^{-j\beta z} + j\beta V^-e^{+j\beta z} = -j\omega LI(z). \tag{1.16}$$

Substitution of (1.13) into (1.16) yields

$$-j\beta V^+e^{-j\beta z} + j\beta V^-e^{+j\beta z} = -j\omega LI^+e^{-j\beta z} + j\omega LI^-e^{+j\beta z}. \tag{1.17}$$

From (1.17), the *characteristic impedance* of a lossless transmission line is then defined as,

$$Z_c = \frac{V^+}{I^+} = \frac{V^-}{I^-} = \frac{\omega L}{\beta} = \sqrt{\frac{L}{C}} = v_p L = \frac{1}{v_p C}. \tag{1.18}$$

It is interesting to note that (1.3)–(1.8), (1.9)–(1.13), and (1.15) are actually in the same form, suggesting that under the TEM-mode operation, the electromegnetic (EM) parameters of a wave-guiding structure can be mapped one-to-one onto the circuit parameters of its transmission-line equivalence. The wave impedance of the parallel-plate waveguide in Fig. 1.1 is in the same form as the characteristic impedance in (1.18), as well.

Table 1.1 Analog between EM parameters in a TEM parallel-plate waveguide and circuit parameters in a uniform transmission line

EM Parameters	Circuit Parameters
E: Electric field intensity (V/m)	*V*: Voltage wave (V)
H: Magnetic field intensity (A/m)	*I*: Current wave (A)
ε: Permittivity (F/m)	*C*: Capacitance per meter (F/m)
μ: Permeability (H/m)	*L*: Inductance per meter (H/m)

Table 1.1 summarizes the analog between the EM parameters of a TEM wave-guiding structure and the circuit parameters of a lossless transmission line. The mapping holds exactly for TEM transmission lines and approximately for quasi-TEM ones. To simplify the design procedure, hereafter we will use the scalar circuit parameters (V, I, L, C) to analyze the propagation characteristics of any kind of TEM/quasi-TEM transmission lines.

1.2.2 Keys to Miniaturization

In Sec. 1.2.1, the wave equation and general solution of a TEM transmission line are derived in terms of the field parameters (E, H, μ, ε) as well as circuit parameters (V, I, L, C) at the same time. In this section, we further introduce the slow wave factor as a figure of merit for judging the circuit miniaturization capability of a given wave-guiding structure.

First of all, recall the guided wavenumber can be expressed in terms of both EM and circuit parameters as

$$\beta_g = \frac{\omega}{v_p} = \omega\sqrt{LC} = \omega\sqrt{\mu\varepsilon} = \omega\sqrt{\mu_r\mu_o\varepsilon_r\varepsilon_o}. \tag{1.19}$$

The free space wavenumber, or the phase constant of a wave propagated in free space, is

$$\beta_o = \omega\sqrt{\mu_o\varepsilon_o}. \tag{1.20}$$

The *slow wave factor* is defined as the ratio of the guided wavenumber to free space wavenumber as

$$SWF = \frac{\beta_g}{\beta_o} = \frac{\lambda_o}{\lambda_g} = \sqrt{\mu_r\varepsilon_r} = c\sqrt{LC}. \tag{1.21}$$

c is the speed of light in vacuum. The slow wave factor is a measure of how good a wave-guiding structure can be used for circuit miniaturization.

Meanwhile, a section of transmission line is commonly expressed in terms of its *electrical length* at the operating frequency as

$$\theta = \beta_g l. \tag{1.22}$$

l is the physical length of the line section. From (1.22), for a given electrical length, increasing the guided wavenumber (β_g) becomes the key factor to reduce the required physical length of a transmission line. Choosing a material media with a higher ε_r or μ_r is a possible way to reduce the physical length with a larger β_g. However, it is likely at the expense of higher fabrication cost. Alternatively, using synthesized transmission lines in accordance with (1.19) and (1.22) paves another way for circuit miniaturization by simultaneously increasing the per-unit-length inductance and capacitance of that line. Here, the *synthesized transmission line* is referred to as any microwave lumped/quasi-lumped network that can be electrically equivalent to a section of uniform transmission line over a frequency band of interest.

A further thought on developing a synthesized transmission line is: in a practical circuit, how can we fulfill the goal by simultaneously increasing the inductance and capacitance of a line? The answer is quite straightforward: "*just follow the fundamental physical rules*." An extra current path always generates additional magnetic fields and, hence, the inductance. The charge accumulation between electrodes, in the meantime, results in extra capacitive loadings. Accordingly, using a meander or spiral high-impedance line is an effective way to provide more current paths or higher current density to increase the per-unit-length inductance in a real design. Meanwhile, adding parallel-plate or interdigital capacitor is a good way to boost the capacitance of the host transmission line. When one attempts to adjust the per-unit-length inductance and capacitance for raising the slow wave factor, it is interesting to note that the characteristic impedance of the line can be controlled at the same time using (1.18) within a reasonable range, say, 20–120 Ω.

Quite a few synthesized transmission lines (or the so-call artificial transmission lines) are summarized and listed in [7]. Some of them are redrawn in Fig. 1.3 for easy reference [8–16]. They are all designed based on alternatively connected series inductance and shunt capacitance with or without periodicity. This sort of synthesized lines is right-handed in nature with lowpass responses. Readers are encouraged to find clues on how the line inductance and capacitance in the examples are boosted. Following the same rule, the readers can develop new and creative structures on their own. In fact, the number of layout patterns of a synthesized transmission line with given electrical properties can be extended to infinity!

Finally, although the aforementioned discussion is restricted to lossless transmission lines, similar statements hold true for a low-loss one. The only difference is the introduction of the

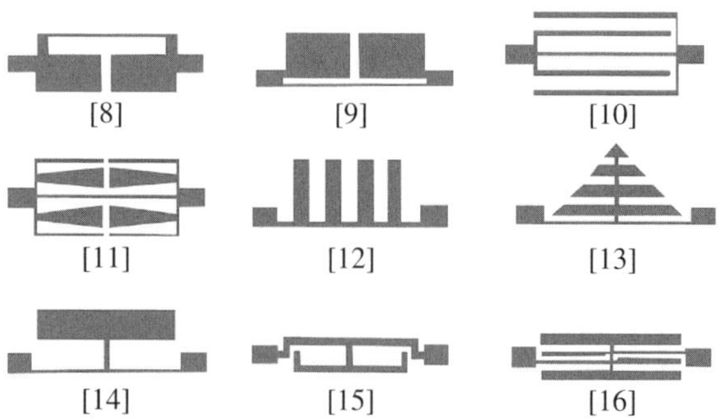

Figure 1.3 Typical slow-wave synthesized transmission lines in open literature

attenuation constant (α), which results in a *complex propagation constant* ($\gamma = \alpha + j\beta$) with a small real part representing the loss. The quality factor (Q) is evaluated by

$$Q = \frac{\beta}{2\alpha}. \tag{1.23}$$

In the following section, the common ways to analyze a periodic or non-periodic synthesized transmission line are introduced as the basis of the entire book.

1.3 Analysis of Synthesized Transmission Lines

Synthesized transmission lines can be realized with or without periodicity. While the non-periodic lines can be dealt with using simple transmission line equivalence, the periodic lines are in general analyzed by the *Bloch theorem* [17]. In this section, the general analysis procedures for periodic and non-periodic synthesized transmission lines are introduced in sequence as follows.

1.3.1 Bloch Theorem and Characterization of a Periodic Synthesized Transmission Line

Figure 1.4 shows a typical periodically-loaded synthesized transmission line. It comprises a uniform transmission line of (Z_c) (*the host line*) periodically loaded by lumped or quasi-lumped shunt elements (*jb*). The line is composed of infinite elements in cascade, with each element termed as a *unit cell*. The voltage and current waves on the *nth* and $(n+1)th$ nodes are related by the *ABCD* matrix of the unit cell as

$$\begin{bmatrix} V_n \\ I_n \end{bmatrix} = \begin{bmatrix} A & B \\ C & D \end{bmatrix} \begin{bmatrix} V_{n+1} \\ I_{n+1} \end{bmatrix}. \tag{1.24}$$

Let the unit cell have a periodicity of p and the complex propagation constant equal to γ, we have

$$\begin{cases} V_{n+1} = V_n \cdot e^{-\gamma p} \\ I_{n+1} = I_n \cdot e^{-\gamma p} \end{cases}. \tag{1.25}$$

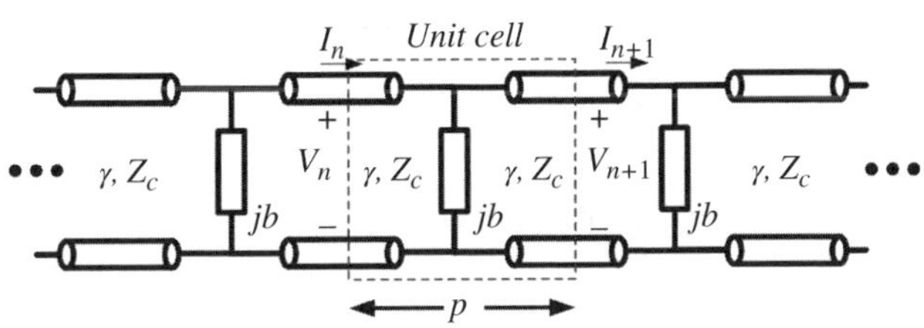

Figure 1.4 Typical periodically-loaded synthesized transmission line

Substitution of (1.25) into (1.24) yields

$$\begin{bmatrix} V_n \\ I_n \end{bmatrix} = \begin{bmatrix} A & B \\ C & D \end{bmatrix} \begin{bmatrix} V_{n+1} \\ I_{n+1} \end{bmatrix} = \begin{bmatrix} V_{n+1} \cdot e^{\gamma p} \\ I_{n+1} \cdot e^{\gamma p} \end{bmatrix}. \tag{1.26}$$

or

$$\begin{bmatrix} A - e^{\gamma p} & B \\ C & D - e^{\gamma p} \end{bmatrix} \begin{bmatrix} V_{n+1} \\ I_{n+1} \end{bmatrix} = 0. \tag{1.27}$$

For a nontrivial solution of (V_{n+1}, I_{n+1}), the determinant of the matrix must be zero, which leads to

$$AD + e^{2\gamma p} - (A + D)e^{\gamma p} - BC = 0. \tag{1.28}$$

Since the unit cell is a reciprocal network, we have $AD - BC = 1$. Thus,

$$e^{2\gamma p} - (A + D)e^{\gamma p} + 1 = 0. \tag{1.29}$$

From (1.29), the propagation constant can be expressed in terms of the *ABCD* matrix as

$$\cosh(\gamma p) = \frac{A + D}{2}. \tag{1.30}$$

In lossless case with $\alpha = 0$, (1.30) is reduced to

$$\cos(\beta p) = \frac{A + D}{2}. \tag{1.31}$$

In a periodic synthesized transmission line, the phase constant of the line can be solved from the *ABCD* matrix of the unit cell with a given periodicity p (i.e., physical length) in accordance with (1.31). In the meantime, the characteristic impedance of the unit cell can be defined as

$$Z_B = \frac{V_{n+1}}{I_{n+1}}. \tag{1.32}$$

This impedance is also called the *Bloch impedance*. From (1.27), we have

$$\left(A - e^{\gamma p}\right)V_{n+1} + BI_{n+1} = 0. \tag{1.33}$$

Substitution of (1.33) into (1.32), the Bloch impedance is

$$Z_B = \frac{-B}{A - e^{\gamma p}}. \tag{1.34}$$

From (1.29), $e^{\gamma p}$ can be solved as

$$e^{\gamma p} = \frac{(A+D) \pm \sqrt{(A+D)^2 - 4}}{2}. \tag{1.35}$$

The Bloch impedance in terms of the *ABCD* matrix is therefore

$$Z_B^{\pm} = \frac{-2B}{A - D \mp \sqrt{(A+D)^2 - 4}}. \tag{1.36}$$

If the unit cell is a symmetric network such as the case in Fig. 1.4, we have $A = D$ and the Bloch impedance can be simplified as

$$Z_B^{\pm} = \frac{\pm B}{\sqrt{A^2 - 1}}. \tag{1.37}$$

Equations (1.31) and (1.37) are the synthesis basis of all sort of one-dimensional periodic structures. They will be used intensively in Chapter 6 for analyzing a periodic synthesized transmission line with two-dimensional routing capability.

1.3.2 Characterization of a Non-Periodic Synthesized Transmission Line

The characterization of a non-periodic synthesized transmission line starts from equating the *ABCD* matrix of the network under development to that of a uniform transmission line with a given characteristic impedance and electrical length, say, Z_c and θ. The equivalences are illustrated in Fig. 1.5. Here, three-element networks, i.e. π-/T-models, are analyzed as an illustration.

The *ABCD* matrices of the π-/T-models are:

$$\begin{bmatrix} A & B \\ C & D \end{bmatrix}_{STL,\pi} = \begin{bmatrix} 1 & 0 \\ Y & 1 \end{bmatrix} \begin{bmatrix} 1 & Z \\ 0 & 1 \end{bmatrix} \begin{bmatrix} 1 & 0 \\ Y & 1 \end{bmatrix} = \begin{bmatrix} 1+ZY & Z \\ Y(2+ZY) & 1+ZY \end{bmatrix}, \tag{1.38a}$$

$$\begin{bmatrix} A & B \\ C & D \end{bmatrix}_{STL,T} = \begin{bmatrix} 1 & Z \\ 0 & 1 \end{bmatrix} \begin{bmatrix} 1 & 0 \\ Y & 1 \end{bmatrix} \begin{bmatrix} 1 & Z \\ 0 & 1 \end{bmatrix} = \begin{bmatrix} 1+ZY & Z(2+ZY) \\ Y & 1+ZY \end{bmatrix}. \tag{1.38b}$$

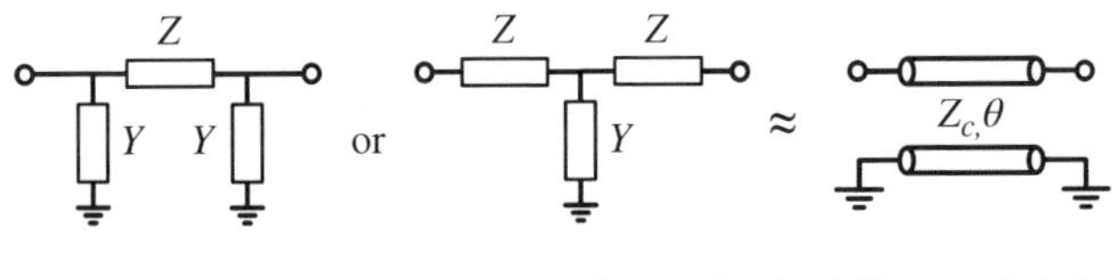

Figure 1.5 Three-element networks as non-periodic synthesized lines and their transmission line equivalence

The *ABCD* matrix of a uniform transmission line is given by

$$\begin{bmatrix} A & B \\ C & D \end{bmatrix}_{TLine} = \begin{bmatrix} \cos\theta & jZ_c \sin\theta \\ jY_c \sin\theta & \cos\theta \end{bmatrix}. \tag{1.39}$$

By equating the corresponding entries in both matrices, one gets;

$$Z_{STL,\pi} = \sqrt{\frac{B}{C}} = \sqrt{\frac{Z}{Y(2+ZY)}} = Z_c, \tag{1.40a}$$

$$Z_{STL,T} = \sqrt{\frac{B}{C}} = \sqrt{\frac{Z(2+ZY)}{Y}} = Z_c, \tag{1.40b}$$

$$\theta_{STL} = \cos^{-1}(1+ZY) = \theta_c. \tag{1.41}$$

At the targeted frequency, Z_{STL} and θ_{STL} are the equivalent characteristic impedance and electrical length of the three-element network as a non-periodic synthesized transmission line. Note that they are a function of frequency and might be equivalent to a uniform transmission line only on a narrowband basis. Similar analysis can be applied to develop a non-periodic synthesized transmission line with an arbitrary network topology, as long as the network is symmetric in configuration. Readers may find this technique repeatedly throughout the book. Last but not the least, although in general the *ABCD* matrix is the most intuitive and straightforward way to complete the design, the analysis could be started with the *Z* or *Y* matrix, as well.

1.3.3 Extraction of Line Parameters from S-Parameters

The propagation parameters of a synthesized transmission line, without or with periodicity, can also be extracted from its two-port *S*-parameters. To illustrate this, the *ABCD* matrix of a general lossy transmission line is first expressed as,

$$\begin{bmatrix} A & B \\ C & D \end{bmatrix} = \begin{bmatrix} \cosh\gamma l & Z_c \sinh\gamma l \\ Y_c \sinh\gamma l & \cosh\gamma l \end{bmatrix}. \tag{1.42}$$

l is the physical length of the line. The relationships between the *ABCD* and *S* matrices are known as

$$A = (1+S_{11}-S_{22}-\Delta S)/(2S_{21}) \tag{1.43a}$$

$$B = (1+S_{11}+S_{22}+\Delta S)Z_o/(2S_{21}) \tag{1.43b}$$

$$C = (1-S_{11}-S_{22}+\Delta S)/(2S_{21}Z_o) \tag{1.43c}$$

$$D = (1-S_{11}+S_{22}-\Delta S)/(2S_{21}) \tag{1.43d}$$

where $\Delta S = S_{11}S_{22} - S_{21}S_{12}$ and Z_0 is the system reference impedance.

Given the fact that the lossy transmission line is symmetric and reciprocal, the complex characteristic impedance (Z_c) and propagation constant (γ_g) are, after some manipulation, extracted from (1.42) and (1.43) as [18]:

$$e^{\gamma_g l} = \frac{1 - S_{11}^2 + S_{21}^2 + \sqrt{\left(1 + S_{11}^2 - S_{21}^2\right)^2 - \left(2S_{11}\right)^2}}{2S_{21}}, \tag{1.44}$$

$$Z_c = Z_o \sqrt{\frac{\left(1 + S_{11}\right)^2 - S_{21}^2}{\left(1 - S_{11}\right)^2 - S_{21}^2}}. \tag{1.45}$$

For a synthesized transmission line, the characteristic impedance and propagation constant can be characterized without much effort by using (1.44) and (1.45) as long as its two-port *S*-parameters are specified by simulation or measurement in advance. The procedure has been widely applied to verify the performance of a synthesized line after the synthesis or integration process.

1.4 Lumped and Quasi-Lumped Approaches

In this and the following sections, classical design approaches are introduced successively to show how a synthesized transmission line can be realized and analyzed. Since the 3-dB branch-line coupler, or the quadrature hybrid, is one of the most common building blocks in microwave systems, it will be adopted as the design example in the following sections for comparison. For each design approach, a branch-line coupler realized by ideal components is first simulated, followed by a real example reported in the literature. The readers may compare the responses of the couplers to learn more about the tradeoff in developing a synthesized transmission line with given specifications.

1.4.1 Lumped Networks

Lumped networks are unambiguously the most intuitive way to fulfill non-periodic synthesized transmission lines [19–21]. Figure 1.6 illustrates a typical synthesized line implemented by lumped elements. It is a π-network comprising two shunt capacitors and a series inductor. By using the technique in Sec. 1.3.2, the *ABCD* matrix of the π-network is

$$\begin{bmatrix} 1 & 0 \\ j\omega C & 1 \end{bmatrix} \begin{bmatrix} 1 & j\omega L \\ 0 & 1 \end{bmatrix} \begin{bmatrix} 1 & 0 \\ j\omega C & 1 \end{bmatrix} = \begin{bmatrix} 1 - \omega^2 LC & j\omega L \\ j\omega C\left(2 - \omega^2 LC\right) & 1 - \omega^2 LC \end{bmatrix}. \tag{1.46}$$

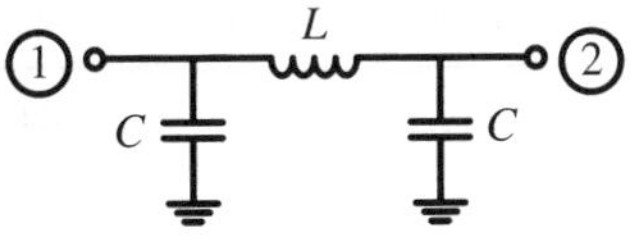

Figure 1.6 Non-periodic synthesized transmission line using a lumped π-network

Meanwhile, the *ABCD* matrix of a lossless uniform transmission line (Z_c, θ) is

$$\begin{bmatrix} A & B \\ C & D \end{bmatrix} = \begin{bmatrix} \cos\theta & jZ_c \sin\theta \\ jY_c \sin\theta & \cos\theta \end{bmatrix}. \tag{1.47}$$

By equating the *ABCD* matrices of the two, the design equations are simplified to

$$\cos\theta = 1 - \omega^2 LC \tag{1.48}$$

$$jZ_c \sin\theta = j\omega L. \tag{1.49}$$

Specifically, the required inductance and capacitance values for a quarter-wavelength synthesized line, with $\theta = \pi/2$, are

$$L = \frac{Z_c}{\omega} \tag{1.50}$$

$$C = \frac{1}{\omega^2 L}. \tag{1.51}$$

An important observation of this sort of synthesized transmission lines is that, unlike a uniform TEM line, the lumped network is lowpass in nature with a cutoff frequency. The 3-dB cutoff frequency can be estimated by

$$f_{cutoff} = \frac{2}{2\pi\sqrt{LC}}. \tag{1.52}$$

In a real design, to preserve phase linearity and non-dispersive characteristic impedance, one should keep in mind that the cutoff frequency of the network should be allocated far away from the targeted operating band to avoid undesired performance degradation.

Using (1.50)–(1.51), the design parameters of a lumped 3-dB quadrature hybrid are summarized in Table 1.2. The center operating frequency for demonstration is set as 1 GHz. The circuit topology is shown in Fig. 1.7(a). The quadrature hybrid consists of pairs of 50- and 35-Ω quarter-wavelength synthesized transmission lines. The simulated *S*-parameters and phase difference between coupled and thru ports are illustrated in Fig. 1.8. The response is somewhat asymmetric with respect to the center frequency, a result of the dispersive nature of the line equivalence. In general, the coupler developed by using the lumped approach has a

Table 1.2 Design parameters of a 1-GHz lumped branch-line coupler

Z_c(Ω)	Frequency (GHz)	$L_{1,2}$ (nH)	$C_{1,2}$ (pF)
50	1	7.958	3.183
35.4	1	5.626	4.502

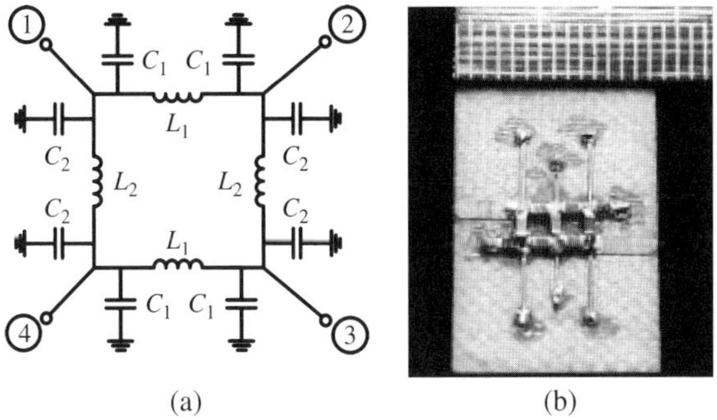

Figure 1.7 Lumped 3-dB branch-line coupler: (a) circuit topology and (b) a design example. Source: Chiang 2001 [22]. Reproduced with permission of IEEE

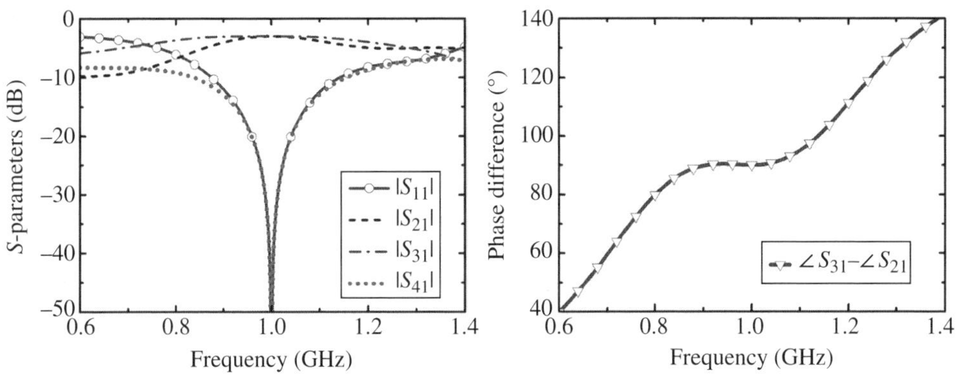

Figure 1.8 S-parameters and phase difference of the lumped coupler

narrower bandwidth than its conventional counterpart. The reduced operating bandwidth can be attributed to the fact that the synthesis equations (1.50)–(1.51) are exact only at the center frequency. Figure 1.7(b) shows the layout of an alternative design [22]. It is realized by lumped π-networks composed of series capacitors and shunt inductors. The alternative network is high-pass or left-handed in nature, but the formulation and line equivalence are the same. Readers may refer to Sec. 1.7 for more information on the circuit realization of metamaterials.

Commercial SMD (surface-mounted device) components are frequently used to fulfill lumped couplers on the printed circuit board (PCB); in the IC implementation, the lumped inductors are generally realized on the thick metal layers in meander or spiral shape. The capacitors, in the meantime, can be implemented by metal-oxide-metal (MOM) or metal-insulator-metal (MIM) structures. Ideally, synthesized transmission lines based on the lumped approach could achieve a very high slow wave factor and hence promising miniaturization capability. Nevertheless, in reality, the parasitic loss, or the Q-factor, ultimately sets an upper bound of the achievable size reduction ratio of the design. In addition, the design flexibility may be limited since the off-the-shelf component values may be restricted.

1.4.2 Shunt-Stub Loaded Lines

In addition to the lumped approach, the shunt-stub loaded synthesized transmission lines have received considerable attention [15, 23], as well. A uniform quarter-wavelength line and its open-stub loaded equivalence using the T-model are shown in Fig. 1.9. Following the same procedure in Sec. 1.3.2, the *ABCD* matrix of the quarter-wavelength synthesized line should be identical to that of the uniform one as

$$\begin{bmatrix} \cos\theta_1 & jZ_1\sin\theta_1 \\ jY_1\sin\theta_1 & \cos\theta_1 \end{bmatrix}\begin{bmatrix} 1 & 0 \\ jY_2\tan\theta_2 & 1 \end{bmatrix}\begin{bmatrix} \cos\theta_1 & jZ_1\sin\theta_1 \\ jY_1\sin\theta_1 & \cos\theta_1 \end{bmatrix} = \begin{bmatrix} 0 & jZ_c \\ jY_c & 0 \end{bmatrix}. \tag{1.53}$$

By equating the entries in the matrices, the synthesis equations for a shunt-stub loaded synthesized line can be derived from (1.53) and summarized as

$$Z_1 = \frac{Z_c}{\tan\theta_1} \tag{1.54}$$

$$Y_2\tan\theta_2 = \frac{2}{Z_1\tan 2\theta_1}. \tag{1.55}$$

Since there are only two equations but four unknowns (Z_1, Z_2, θ_1, θ_2), the designer has two degrees of freedom. A variety of circuit layouts, by selecting different variables as the initial parameters, are therefore reported in the literature.

The duality of the T-model, i.e. the π-model in Fig. 1.10, can be equivalent to a quarter-wavelength uniform transmission line, as well. Following the same analysis, the design equations are summarized as

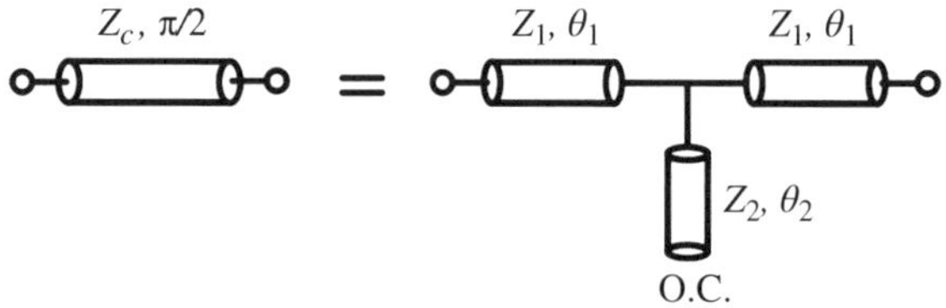

Figure 1.9 A uniform quarter-wavelength line and its equivalent shunt-stub loaded synthesized line using the T-model

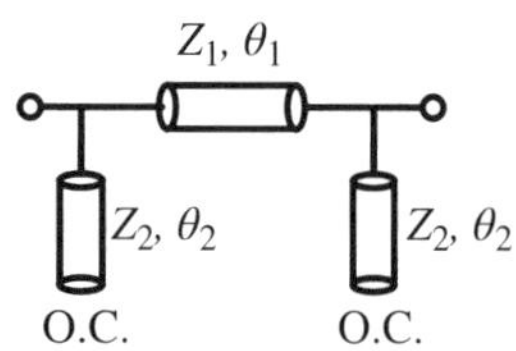

Figure 1.10 A shunt-stub loaded synthesized line based on the π-model

$$Z_1 = \frac{Z_c}{\sin\theta_1} \tag{1.56}$$

$$Y_2 \tan\theta_2 = \frac{\cos\theta_1}{Z_c}. \tag{1.57}$$

A 3-dB branch-line coupler is designed as an example. The operating frequency is also 1 GHz for easy comparison. The T-model in Fig. 1.9 is adopted as the building block. Let $\theta_1 = 30°$ and $Z_2 = 30\ \Omega$ for the 35- and 50-Ω synthesized lines, one can use (1.54)–(1.55) to calculate the remaining parameters as listed in Table 1.3. Figure 1.11 illustrates the S-parameters of the coupler; the phase difference between the thru and coupled ports is shown at the same time. The S-parameters are asymmetric with respect to the center frequency since the cutoff frequency of the non-periodic synthesized line, also low-pass in nature, may be too close to the band of concern. In addition, the network equivalence in (1.54) and (1.55) is on a narrow-band basis, which may account for the discrepancy, as well.

For efficient usage of the circuit area, the open stub in the synthesized line can be further split into multiple stubs. Figure 1.12 shows an example by replacing the original stub with two open stubs connected in parallel [23]. The design equation is straightforward as

$$\frac{Z_a \tan\theta - Z_c \tan\theta_a}{Z_a^2 \tan\theta \tan\theta_a + Z_c Z_a} = Y_1 \tan\theta_1 + Y_2 \tan\theta_2. \tag{1.58}$$

Table 1.3 Design parameters of a 1-GHz shunt-stub branch-line coupler

Z_c (Ω)	Frequency (GHz)	Z_1 (Ω)	θ_1(°)	Z_2 (Ω)	θ_2(°)
50	1	86.60	30	30	21.8
35.4	1	61.23	30	30	29.5

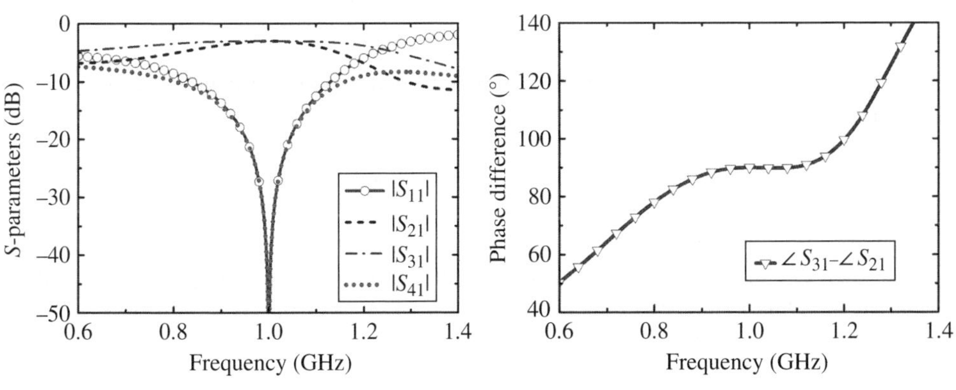

Figure 1.11 S-parameters and phase difference of the 3-dB coupler realized by loaded shunt stubs

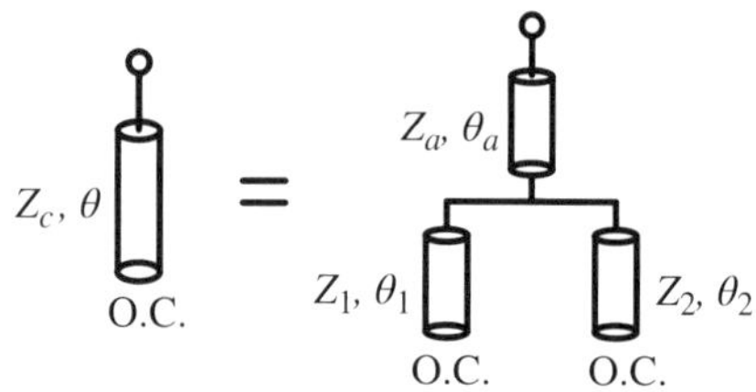

Figure 1.12 Splitting a single stub into dual open stubs in parallel connection

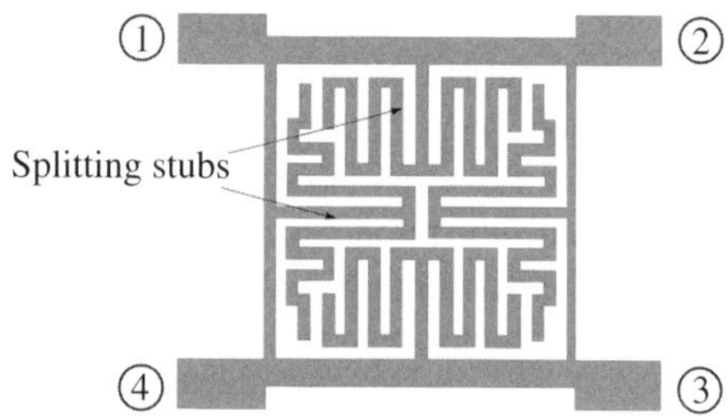

Figure 1.13 Typical layout of a shunt-stub loaded coupler using split open stubs. Source: Tang 2007 [23]. Reproduced with permission of IEEE

Figure 1.13 shows the typical layout of a 3-dB quadrature hybrid using split stubs. The stubs are meandered to fill out the inner space of the coupler. It is 23.5% the size of a conventional design. Splitting a single stub into triple or even more stubs for circuit miniaturization is also possible, yet the parasitic coupling between adjacent lines makes the design highly dependent on an EM simulator to optimize the layout for good responses.

The ways to implement synthesized transmission lines using non-periodic lumped/quasi-lumped networks are included but not limited to the aforementioned approaches. For example, the bridged T-coil in [24] is a very elegant way to fulfill a very compact design. The readers can find more design examples and new ideas in Chapter 2 about PCB designs and in Chapter 5 about on-chip realization.

1.5 One-Dimensional Periodic Structures

Periodic structures are also a popular way to realize synthesized transmission lines for circuit miniaturization [25, 26]. In this section, one-dimensional periodic structures will be analyzed using the Bloch wave analysis in Sec. 1.3.1; the two-dimensional ones, in the meantime, will be covered in Sec. 1.6.

In general, a one-dimensional periodic synthesized line is composed of periodically-loaded inductors and capacitors, and can be directly mapped to the distributed transmission-line model in Fig. 1.2. Figure 1.14(a) shows a typical one-dimensional periodic synthesized line. The unit cell can be represented by either a T- or π-model. As an example, the T-model is discussed in terms of the two-port z-parameters in Fig. 1.14(b). According to (1.30) and Fig. 1.14(b), the propagation constant can be expressed as

$$\cosh(\gamma p) = \frac{A+D}{2} = \frac{Z_{11}+Z_{22}}{2Z_{12}}. \tag{1.59}$$

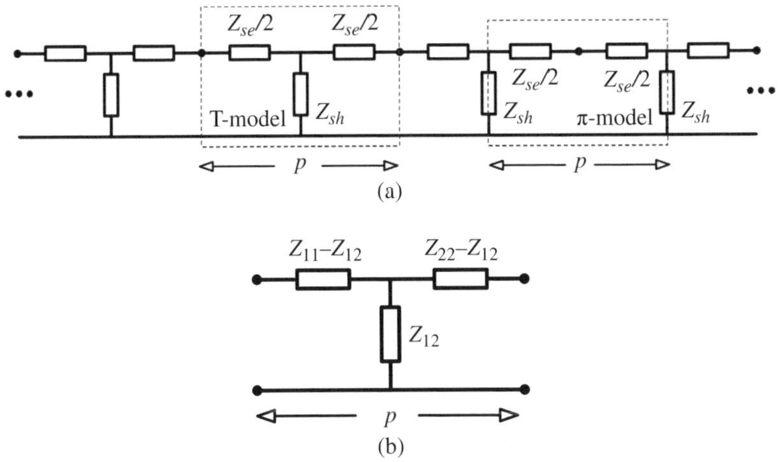

Figure 1.14 Typical one-dimensional periodic synthesized transmission line: (a) infinite unit cells in cascade and (b) two-port *Z*-parameters of the unit cell using the T-model

Since the unit cell is much smaller than the guided wavelength ($p << \lambda_g$), the hyper-trigonometric function can be simplified to:

$$\cosh(\gamma p) \cong 1 + \frac{(\gamma p)^2}{2}. \tag{1.60}$$

Substitution of (1.60) into (1.59) gets,

$$(\gamma p)^2 = \frac{Z_{11} + Z_{22} - 2Z_{12}}{Z_{12}} \equiv \frac{Z_{se}}{Z_{sh}}. \tag{1.61}$$

Here, Z_{se} is the direct sum of the series impedances of the T-network, and Z_{sh} is the impedance of the shunt stub. The propagation constant is therefore characterized by

$$\gamma = \alpha + j\beta = \frac{1}{p}\sqrt{\frac{Z_{se}}{Z_{sh}}}, \tag{1.62}$$

and the Bloch impedance of the periodic synthesized line is

$$Z_B = \sqrt{Z_{se} Z_{sh}}. \tag{1.63}$$

Equations (1.62)–(1.63) are the design basis of a one-dimensional periodic synthesized transmission line.

Consider again the 1-GHz branch-line coupler. Based on the T-model, a lossless periodic *LC* network is shown in Fig. 1.15. The series impedance (Z_{se}) and shunt impedance (Z_{sh}) are, respectively,

$$Z_{se} = j\omega L \tag{1.64a}$$

$$Z_{sh} = 1 / j\omega C. \tag{1.64b}$$

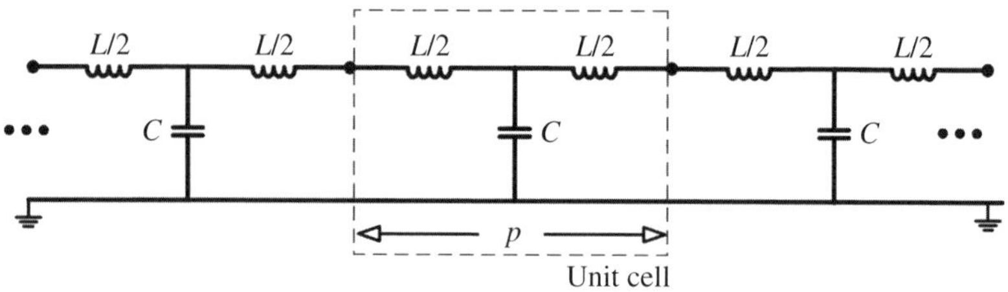

Figure 1.15 A periodic synthesized transmission line using T-model unit cells

Table 1.4 Parameters of a 3-dB branch-line coupler by periodic lumped *LC* unit cells

$Z_c(\Omega)$	Frequency (GHz)	N_{UC}	C (pF)	L (nH)
50	1	30	0.1667	0.4167
35.4	1	25	0.2829	0.3535

Substitution of (1.64) into (1.62) and (1.63) yields,

$$j\beta = \frac{1}{p}\cdot j\omega\sqrt{LC} \tag{1.65}$$

$$Z_c = \sqrt{\frac{L}{C}}. \tag{1.66}$$

If the number of unit cells in a quarter-wavelength line is specified as N_{UC}, the design equations for the inductance and capacitance values in each unit cell become

$$C = \frac{\pi/2}{N_{UC}}\left(\frac{1}{\omega Z_c}\right) \tag{1.67}$$

$$L = Z_c^2\cdot C. \tag{1.68}$$

Using (1.67) and (1.68), the parameters for realizing a 1-GHz branch-line coupler are summarized in Table 1.4. The numbers of unit cells in the 35-Ω and 50-Ω periodic synthesized lines are 25 and 30, respectively. Figure 1.16 shows its corresponding *S*-parameters and phase difference. Since the periodic structure is more similar in topology to a uniform transmission line, the operating bandwidth of the coupler is very similar to that of a conventional design. The cutoff frequency of the synthesized line also reflects the same trend. For this specific design, the cut-off frequency of the unit cell is as high as 18.9 GHz, making the responses very symmetric with respect to the center frequency.

In a real design, the inductors and capacitors in the one-dimensional periodic structure are commonly realized by high-impedance lines and shunt open stubs, respectively. Although the design methodology is similar, one should be aware that since there are only two requisite

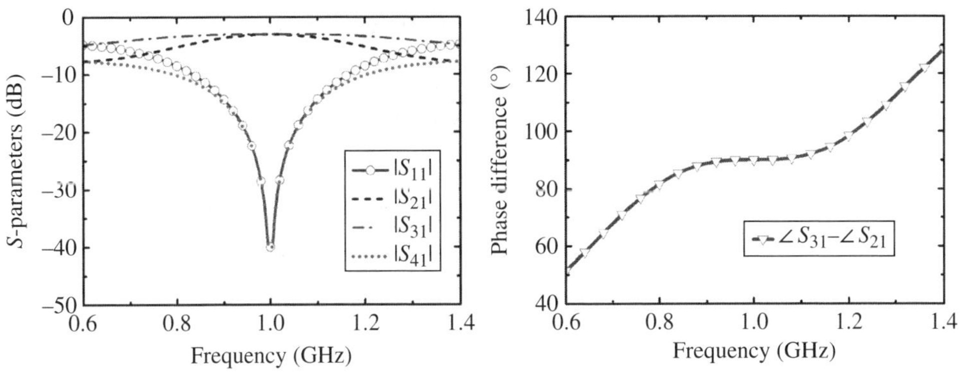

Figure 1.16 *S*-parameters and phase difference of the 3-dB coupler by one-dimensional periodic synthesized lines with lumped *LC* sections

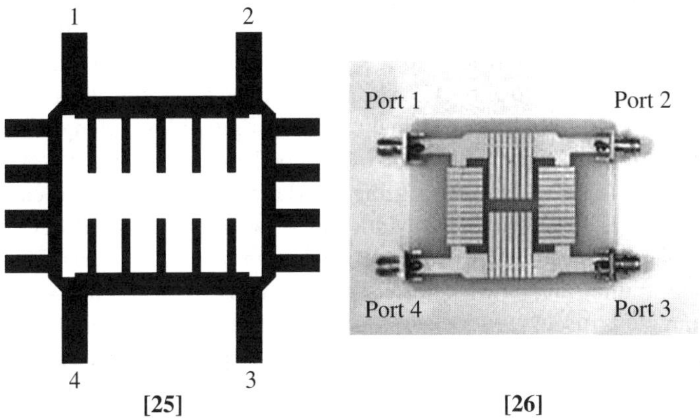

Figure 1.17 Design examples of quadrature hybrids using periodic synthesized transmission lines. Source: Eccleston 2003 [25]; Sun 2005 [26]. Reproduced with permission of IEEE

conditions (Z_c, θ) but four unknowns (characteristic impedances and electrical lengths of the series/shunt lines), a number of topologies with equal electrical properties can be developed for layout optimization and circuit miniaturization. Typical examples are shown in Fig. 1.17. The number of unit cells in a quarter-wavelength line varies from 4 to 12. Readers may also use the same procedure to develop their own periodic synthesized transmission lines with a unique layout arrangement.

1.5.1 Complementary-Conducting-Strip Lines

The complementary-conducting-strip (CCS) line is a one-dimensional periodic synthesized line especially suitable for semiconductor fabrication processes. It can facilitate two-dimensional routing and has been widely used in microwave systems [2, 3, 27–29]. Figure 1.18(a) shows the unit cell of a typical CCS transmission line. The signal trace is composed of a central patch

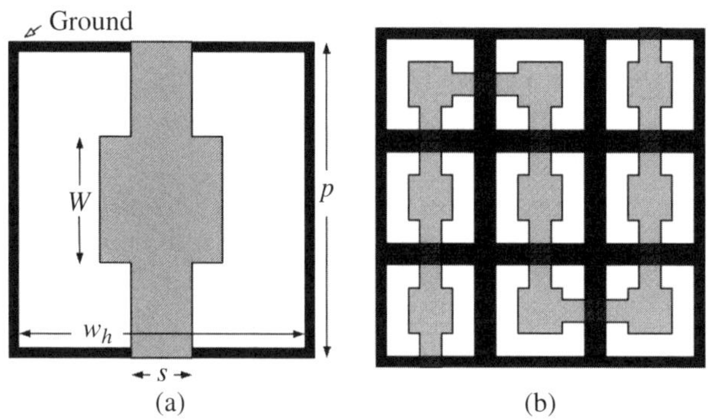

Figure 1.18 CCS synthesized transmission line: (a) unit cell and (b) a meander line

and connection arms. The widths of the patch and connection arm are w and s, respectively. The ground plane is deposited around the patch with a mesh area of $w_h \times w_h$. Since the arms of the CCS cell can be connected to two out of four directions, the CCS line can fulfill two-dimensional routing with very efficient usage of the circuit area for miniaturization.

The Bloch wave analysis in Sec. 1.3.1 can be again applied to develop the line. The periodicity of the unit cell, which should be much smaller than the guided wavelength, is denoted as p. The shunt capacitance is determined by the overlapped area between the ground plane mesh ($w_h \times w_h$) and signal trace (s). Increasing the overlapped area raises the capacitance. The inductance is determined by the patch width (w) as well as the signal trace (s). Since the inductance and capacitance of a CCS cell can be determined independently, one can fulfill a two-dimensional routing with a wide variety of characteristic impedances.

A design example is discussed for clear illustration. In a 1P6M 0.18 μm CMOS process, the design parameters for a CCS unit cell are $p = 30\,\mu m$, $w_h = 29\,\mu m$, $w = 5\,\mu m$, and $s = 3\,\mu m$ [29]. The thicknesses of the signal trace and meshed ground plane are 2.39 and 6.05 μm, respectively. By a cascade connection of nine unit cells, the layout of the meander line is shown in Fig. 1.18(b); the total length is 270 μm. By using an EM simulator, the S-parameters of the CCS sample at 24 GHz are $S_{11} = 0.053 + j0.088$ and $S_{21} = 0.921 - j0.304$. The extracted characteristic impedance from (1.45) is $70.28 - j3.59\ \Omega$. The image part accounts for the conductor and dielectric losses. The slow-wave factor and quality factor at 24 GHz are 2.25 and 7.6, respectively.

The CCS synthesized line is non-dispersive over a wide frequency range. As a rule of thumb, the variation of the characteristic impedance of the meander line in Fig. 1.18(b) is less than 3.6% from 10 to 40 GHz. Its slow wave factor, which is around 2.3, is actually not remarkable at all. Nevertheless, the two-dimensional routing capability makes the CCS line become a promising solution for circuit miniaturization with great layout flexibility.

1.6 Photonic Bandgap Structures

The photonic bandgap (PBG) structure is a real two-dimensional synthesized transmission line. Unlike the previous cases, the periodic patterns are fabricated on the entire ground plane rather than the signal trace [30, 31]. Figure 1.19 illustrates the typical layout of nine PBG unit cells. Each unit cell consists of a square metal pad along with four connecting arms. The ground

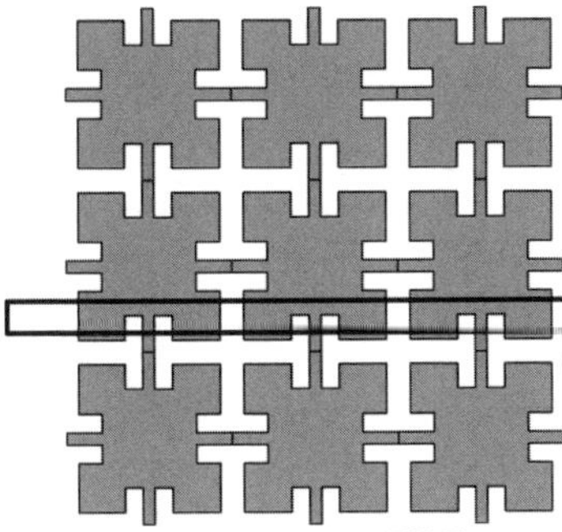

Figure 1.19 A typical PBG structure with an overlaid microstrip line

plane is etched with a periodic pattern, and the signal trace above the ground plane could be any sort of conventional transmission line, such as a microstrip line. While the narrow arms of the unit cell lead to additional inductance, the coupling between nearby pads raises the overall capacitance. The structure hence reveals slow-wave characteristics.

Despite its slow-wave nature, the main purpose of PBG structures is to introduce bandgap phenomenon for mode suppression, system isolation, and so on. During the design phase, a full-wave EM simulator is always required to complete the design since analyzing the PBG structure using simple circuit models is nearly impossible. For interested readers, more information can be found in [30].

1.7 Left-Handed Structures

Left-handed structures, or the metamaterials, are also a popular solution not only for size miniaturization but also for realizing a number of circuits not feasible by conventional righted-handed components. For example, dual-band designs with a non-integer ratio between passbands, beam-switching agility of leaky-wave antennas, and so on are realized and demonstrated [32, 33]. In fact, the concept of metamaterials is not new, and can be traced back to 1967 by a Russian physicist, Victor Veselago [34]. Nevertheless, it was not introduced to the microwave and antenna communities until the late 1990s and early 2000s. After the induction, related research was boosted dramatically, and this has been recognized as one of the main streams in the 2010s. The interested readers may found more information in a number of books such as [35, 36].

As shown in Fig. 1.20(a), a pure left-handed transmission line is a periodic structure consisting of alternatively connected series capacitance and shunt inductance. By using the Bloch wave analysis in Sec. 1.3.1, the propagation characteristics of the left-handed structure can be extracted. Some interesting phenomena, such as negative permittivity and permeability, anti-parallel phase and group velocities, and so on are observed. However in the real implementation, since the series capacitance and shunt inductance are inevitably connected to one another by conventional, or right-handed, transmission lines, a more realistic model named the composite right/left-handed (CRLH) transmission line was proposed and analyzed. Referring to Fig. 1.20(b), the CRLH line consists of periodically connected series *LC* resonator in the serial path and parallel *LC* resonator in the shunt path. While the C_L and L_L are the left-handed components, the C_R and L_R account for the right-handed connecting lines.

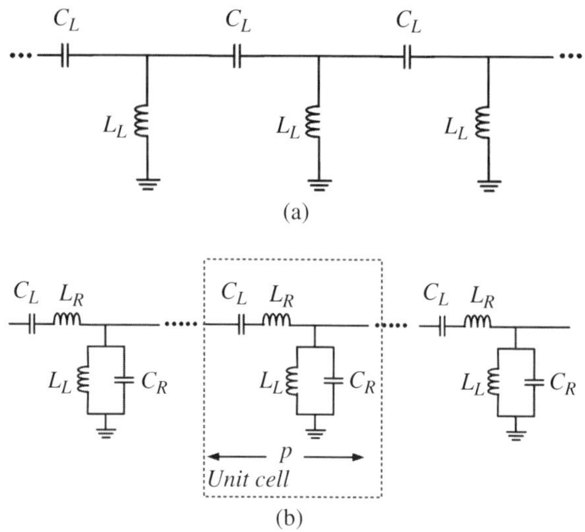

Figure 1.20 Periodic unit cells of (a) a pure left-handed synthesized line and (b) a composite right/left handed (CRLH) transmission line

The propagation constant and characteristic impedance, under a lossless assumption, can be solved from (1.62) and (1.63) as

$$\beta = \frac{1}{p}\cos^{-1}\left[1 - \frac{1}{2}\left(\frac{\omega_L^2}{\omega^2} + \frac{\omega^2}{\omega_R^2} - \frac{\omega_L^2}{\omega_{se}^2} - \frac{\omega_L^2}{\omega_{sh}^2}\right)\right] \tag{1.69}$$

$$Z_c = \sqrt{\frac{L_L}{C_L}} \cdot \sqrt{\frac{\omega^2 / \omega_{se}^2 - 1}{\omega^2 / \omega_{sh}^2 - 1}} \tag{1.70}$$

where

$$\omega_L = \frac{1}{\sqrt{C_L L_L}} \quad \omega_R = \frac{1}{\sqrt{C_R L_R}} \quad \omega_{se} = \frac{1}{\sqrt{C_L L_R}} \quad \omega_{sh} = \frac{1}{\sqrt{C_R L_L}}. \tag{1.71}$$

p is the periodicity. If one further assumes

$$\omega_{se} = \frac{1}{\sqrt{C_L L_R}} = \frac{1}{\sqrt{C_R L_L}} = \omega_{sh}, \tag{1.72}$$

the CRLH line is said to be balanced so that the equations are reduced as

$$\beta = \omega\sqrt{L_R C_R} - \frac{1}{\omega\sqrt{L_L C_L}} \tag{1.73}$$

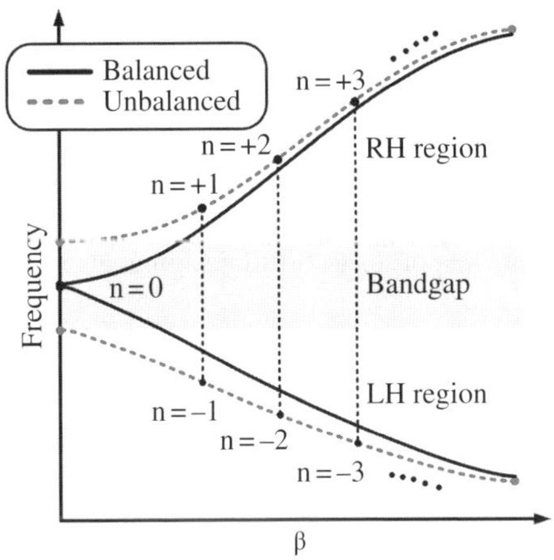

Figure 1.21 Dispersion diagrams of the balanced and unbalanced CRLH lines

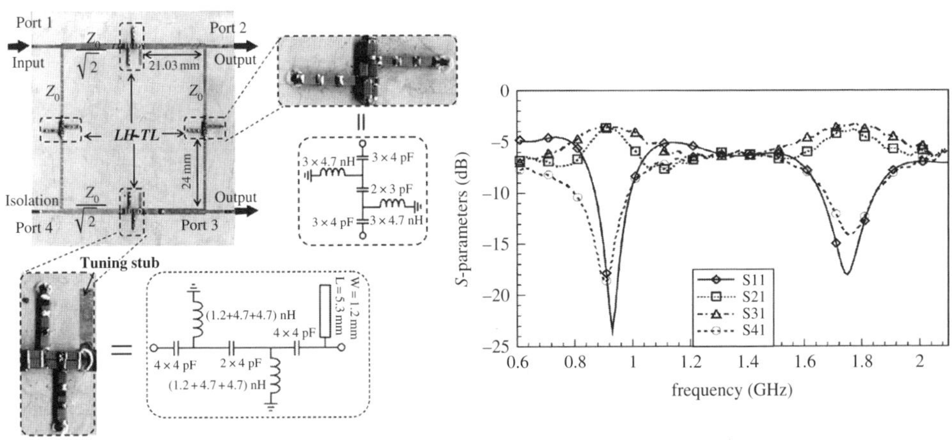

(a) (b)

Figure 1.22 (a) Layout and (b) *S*-parameters (measurement) of a dual-band branch-line coupler using CRLH lines. Source: Lin 2004 [37]. Reproduced with permission of IEEE

$$Z_c = \sqrt{\frac{L_L}{C_L}} = \sqrt{\frac{L_R}{C_R}}. \tag{1.74}$$

Figure 1.21 compares the dispersion diagrams of the balanced and unbalanced CRLH transmission lines. For the unbalanced case, a bandgap between the left-handed and right-handed passbands is clearly observed. With the balanced condition, the bandgap is disappeared and the operating band becomes continuous. A number of positive as well as negative resonance modes, occurring when $\beta p=\pm n\pi$, are observed. They have been widely applied to various applications with distinct circuit properties.

Figure 1.22(a) shows a typical design example, which is a dual-band branch-line coupler [37]. Each CRLH line is realized by a pair of conventional microstrip lines along with two

T-model left-handed unit cells. The ratio between the operating frequency bands, as shown in Fig. 1.22(b), is a non-integer. This feature is not attainable by uniform wave-guiding structures. A left-handed synthesized line without periodicity is also realizable. In this case, the network could be analyzed using the *ABCD* matrix as mentioned in Sec. 1.3.2. Generally, it comprises a number of series capacitors and shunt inductors in cascade and is highpass in nature. A number of left-handed synthesized lines are developed in the following chapters to utilize their unusual frequency responses to fulfill design goals such as multi-mode operation.

References

[1] C. Z. Li and J. S. Lin, *Microwave Noncontact Motion Sensing and Analysis*, 1st edn, John Wiley & Sons, Ltd, Chichester, UK, 2013.

[2] S. Wang, K.-H. Tsai, K.-K. Huang, and M.-S. Kang, S.-X. Li, H.-S. Wu and C.-K. C. Tzuang, "Design of *X*-band RF CMOS transceiver for FMCW monopulse radar," *IEEE Trans. Microw. Theory Techn.*, vol. 57, no. 1, pp. 61–70, Jan. 2009.

[3] H.-S. Wu, C.-W. Wang, and C.-K. C. Tzuang, "CMOS active quasi-circulator with dual transmission gains incorporating feedforward technique at *K*-band," *IEEE Trans. Microw. Theory Techn.*, vol. 58, no. 8, pp. 2084–2091, Aug. 2010.

[4] P. Pursula, T. Karttaavi, M. Kantanen, A. Lamminen, J. Holmberg, M. Lahdes, et al., "60-GHz millimeter-wave identification reader on 90-nm CMOS and LTCC," *IEEE Trans. Microw. Theory Techn.*, vol. 59, no. 4, pp. 1166–1173, Apr. 2011.

[5] J.-L. Kuo, Y.-F. Lu, T.-Y. Huang, Y.-L. Chang, Y.-K. Hsieh, P.-J. Peng, et al., "60-GHz four-element phased-array transmit/receive system-in-package using phase compensation techniques in 65-nm flip-chip CMOS process," *IEEE Trans. Microw. Theory Techn.*, vol. 60, no. 3, pp. 743–756, Mar. 2012.

[6] S.-Y. Kim, O. Inac, C.-Y. Kim, D. Shin, and G. M. Rebeiz, "A 76–84-GHz 16-element phased-array receiver with a chip-level built-in self-test system," *IEEE Trans. Microw. Theory Techn.*, vol. 61, no. 8, pp. 3083–3098, Aug. 2013.

[7] P. Kurgan, J. Filipcewicz, and M. Kitlinski, "Development of a compact microstrip resonant cell aimed at efficient microwave component size reduction," *IET Microwaves, Antennas & Propagation.*, vol. 6, no. 12, pp. 1291–1298, Aug. 2012.

[8] J.-S. Hong and M. J. Lancaster, "Theory and experiment of novel microstrip slow-wave open-loop resonator filters," *IEEE Trans. Microw. Theory Techn.*, vol. 45, no. 12, pp. 2358–2365, Dec. 1997.

[9] C. Karpuz, "Bandstop characteristics of a triangular microstrip slotted patch as an electromagnetic bandgap (EBG)," *Microw. Opt. Technol. Lett.*, vol. 36, no. 3, pp. 149–150, Jan. 2003.

[10] G. Cakir and L. Sevgi, "A design of novel microstrip electromagnetic bandgap (EBG) structure," *Microw. Opt. Technol. Lett.*, vol. 46, no.4, pp. 399–401, Jun. 2005.

[11] M. Hayati and A. Lotfi, "Compact lowpass filter with high and wide rejection in stopband using front coupled tapered CMRC," *Electron. Lett.*, vol. 46, no.12, pp. 846–848, Jun. 2010.

[12] K.W. Eccleston and S.H.M. Ong, "Compact planar microstripline branch-line and rat-race couplers," *IEEE Trans. Microw. Theory Techn.*, vol. 51, no. 10, pp. 2119–2125, Oct. 2003.

[13] J. Wang, B. Z. Wang, Y. X. Guo, L. C. Ong, and S. Xiao, "A compact slow-wave microstrip branch-line coupler with high performance," *IEEE Microw. Wireless Compon. Lett.*, vol. 17, no. 7, pp. 501–503, Jul. 2007.

[14] K. S. Chin, K. M. Lin, Y. H. Wei, T. H. Tseng, and Y. J. Yang, "Compact dual-band branch-line and rat-race couplers with stepped-impedance-stub lines," *IEEE Trans. Microw. Theory Techn.*, vol. 58, no. 5, pp. 1213–1221, May. 2010.

[15] C. H. Tseng and H. J. Chen, "Compact rat-race coupler using shunt-stub-based artificial transmission lines," *IEEE Microw. Wireless. Compon. Lett.*, vol. 18, no. 11, pp. 734–736, Nov. 2008.

[16] J. He, B. Z. Wang, and W. Shao, "Compact power divider embedded with zigzag microstrip slow-wave structures," *Electron. Lett.*, vol. 45, no. 1, pp. 62–63, Jan. 2009.

[17] R. E. Collin, *Foundations for Microwave Engineering*, 2nd edn, New York: McGraw-Hill, 1992.

[18] W.-R. Eisenstadt and Y. Eo, "S-parameter-based IC interconnect transmission line characterization," *IEEE Trans. Comp., Hybrids, Manufact. Technol.*, vol. 15, pp. 483–490, Aug. 1992.

[19] R.-W. Vogel, "Analysis and design of lumped- and lumped-distributed-element directional couplers for MIC and MMIC applications," *IEEE Trans. Microw. Theory Techn.*, vol. 40, no. 2, pp. 253–262, Feb. 1992.
[20] E. Gandini, M. Ettorre, R. Sauleau, and A. Grbic, "A lumped-element unit cell for beam-forming networks and its application to a miniaturized Butler matrix," *IEEE Trans. Microw. Theory Techn.*, vol. 61, no. 4, pp. 1477–1487, Apr. 2013.
[21] B. Cetinoneri, Y.-A. Atesal, and G. M. Rebeiz, "An 8×8 Butler matrix in 0.13-μm CMOS for 5–6-GHz multi-beam applications," *IEEE Trans. Microw. Theory Techn.*, vol. 59, no. 2, pp. 295–301, Feb. 2011.
[22] Y.-C. Chiang and C.-Y. Chen, "Design of a wide-band lumped-element 3-dB quadrature coupler," *IEEE Trans. Microwave Theory Techn.*, vol. 49, no. 3, pp. 476–479, Mar. 2001.
[23] C.-W. Tang and M.-G. Chen, "Synthesizing microstrip branch-line couplers with predetermined compact size and bandwidth," *IEEE Trans. Microw. Theory Techn.*, vol. 55, no. 9, pp. 1926–1934, Sept. 2007.
[24] Y.-S. Lin and J.-H. Lee, "Miniature Butler matrix design using glass-based thin-film integrated passive device technology for 2.5-GHz applications," *IEEE Trans. Microw. Theory Techn.*, vol. 61, no. 7, pp. 2594–2602, July 2013.
[25] K.-W. Eccleston and S.-H. M. Ong, "Compact planar microstripline branch-Line and rat-race couplers," *IEEE Trans. Microw. Theory Techn.*, vol. 51, no. 10, pp. 2119–2125, Oct. 2003.
[26] K.-O. Sun, S.-J. Ho, C.-C. Yen, and D.-Van D. Weide, "A compact branch-line coupler using discontinuous microstrip lines," *IEEE Microw. Wireless Compon. Lett.*, vol. 15, no. 8, pp. 519–520, Aug. 2005.
[27] C.-C. Chen and C.-K. C. Tzuang, "Synthesized quasi-TEM meandered transmission lines for compacted microwave integrated circuits," *IEEE Trans. Microw. Theory Techn.*, vol. 52, no. 6, pp. 1637–1647, Jun. 2004.
[28] M.-J. Chiang, H.-S. Wu, and C.-K. C. Tzuang, "Design of synthesized quasi-TEM transmission line for CMOS compact integrated circuit," *IEEE Trans. Microw. Theory Techn.*, vol. 55, no. 12, pp. 2512–2520, Dec. 2007.
[29] C.-W. Wang, H.-S. Wu, and C.-K. C. Tzuang, "CMOS passive phase shifter with group-delay deviation of 6.3 ps at *K*-Band," *IEEE Trans. Microw. Theory Techn.*, vol. 59, no. 7, pp. 1778–1786, Jul. 2011.
[30] F.-R. Yang, K.-P. Ma, Y. Qian, and Tatsuo Itoh, "A uniplanar compact photonic-bandgap (UC-PBG) structure and its applications for microwave circuits," *IEEE Trans. Microw. Theory Techn.*, vol. 47, no. 8, pp. 1509–1514, Aug. 1999.
[31] F.-R. Yang, Y. Qian, R. Coccioli, and T. Itoh, "A novel low-loss slow-wave microstrip structure," *IEEE Microw. Guided Wave Lett.*, vol. 8, no. 11, pp. 372–374, Nov. 1998.
[32] G. V. Eleftheriades, "Enabling RF/microwave devices using negative refractive-index transmission-line (NRI-TL) metamaterials," *IEEE Antennas Propag. Mag.*, vol. 49, no. 2, pp. 34–51, Apr. 2007.
[33] C. Caloz, T. Itoh, and A. Rennings, "CRLH metamaterial leaky-wave and resonant antennas," *IEEE Antennas Propag. Mag.*, vol. 50, no. 5, pp. 25–39, Oct. 2008.
[34] V. G. Veselago, "The electrodynamics of substances with simultaneously negative value of ε and μ," *Soviet Phys. Uspekhi*, vol. 10, no. 4, pp. 509–514, Jan. 1968.
[35] N. Engheta and R. W. Ziolkowski, *Metamaterial: Physics and Engineering Explorations*, 1st edn, John Wiley & Sons, Ltd, Chichester, UK, 2006.
[36] C. Caloz and T. Itoh, *Electromagnetic Metamaterials: Transmission Line Theory and Microwave Applications*, Wiley-IEEE Press, Chichester, UK, 2005.
[37] I.-H. Lin, M. DeVincentis, C. Caloz, and T. Itoh, "Arbitrary dual-band components using composite right/left-handed transmission lines," *IEEE Trans. Microw. Theory Techn.*, vol. 52, no. 4, pp. 1142–1149, Apr. 2004.

2

Non-Periodic Synthesized Transmission Lines for Circuit Miniaturization

C. W. Wang and T. G. Ma

2.1 Introduction

Non-periodic synthesized transmission lines are realized by quasi-lumped/lumped networks without any periodicity. In this sort of designs, the synthesized line is realized by a network associated with a significant amount of electrical length (i.e., phase delay). It can be directly applied as the building blocks of various passive microwave components. In general, the electrical length of a non-periodic synthesized transmission line ranges from 30° to 90°, but can be even more. The common features of a non-periodic synthesized transmission line include a high slow wave factor, impressive circuit miniaturization capability, and a simple synthesis approach. Nevertheless, they are also known to suffer from dispersive responses and higher parasitic loss. A compromise between the size reduction ratio, power dissipation and phase linearity is always required.

In this chapter, a number of non-periodic synthesized transmission lines, using quasi-lumped/lumped elements in microstrip and coplanar waveguide (CPW) forms on printed circuit boards (PCBs), are introduced using the design principles in Chapter 1. The synthesized microstrip/CPW lines are applied to a variety of microwave circuits with or without coupling between lines. Essential building blocks in the feeding network of a phased array, including the hybrid couplers and Butler matrix, are realized by synthesized microstrip lines. A bandpass filter, based on a synthesized dual-mode ring resonator, is investigated and demonstrated. The ring resonator has more stringent restrictions on the unloaded quality factor of the line used. Coupled synthesized lines using CPW structures or lumped elements are studied and applied to realize directional couplers, Marchand baluns, and reflection-type phase shifters. Finally, the concept of vertical inductors is introduced at the end of the chapter.

Synthesized Transmission Lines: Design, Circuit Implementation, and Phased Array Applications, First Edition.
Tzyh-Ghuang Ma, Chao-Wei Wang, Chi-Hui Lai and Ying-Cheng Tseng.

The non-periodic synthesized lines are the fundamentals of the multi-operational mode transmission lines introduced in the next two chapters.

2.2 Non-Periodic Synthesized Microstrip Lines and Their Applications

2.2.1 Design Details and Propagation Characteristics

A quarter-wavelength synthesized microstrip line using quasi-lumped inductors and capacitors is shown in Fig. 2.1(a) [1]. Its corresponding equivalent lumped model is plotted in Fig. 2.1(b). The synthesized line consists of three stages of quasi-lumped networks, with the first and last stages comprising a parallel connection of a meander-line inductor (L_1, L_3) and parallel-plate capacitors ($C_{s1} \sim C_{s4}$). The intermediate stage is an LC tank using a meander-line inductor (L_2) and interdigital capacitors (C_1) for introducing a transmission zero to suppress the harmonics. The meander-line inductors and interdigital capacitors inevitably introduce parasitic shunt capacitances. Their values are denoted as $C_{l1} \sim C_{l4}$ and $C_{p1} \sim C_{p4}$, respectively.

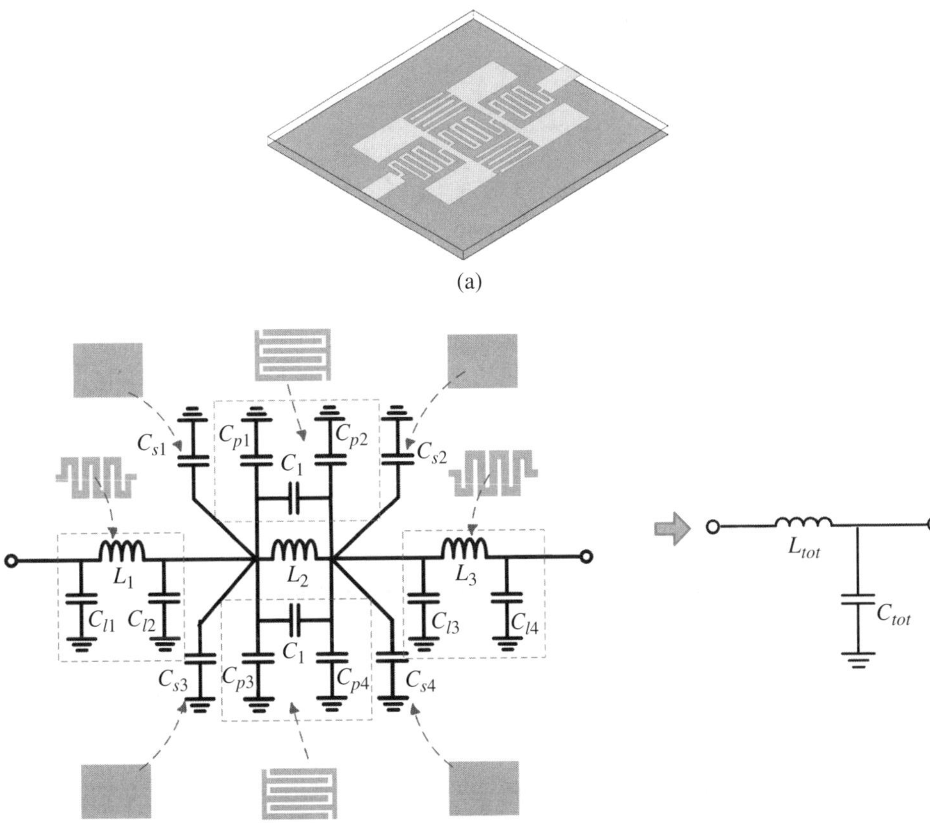

Figure 2.1 Non-periodic synthesized microstrip line: (a) circuit layout and (b) equivalent lumped circuit model. Source: Wang 2007 [1]. Reproduced with permission of IEEE

Since, from (1.21), the slow wave factor as well as size reduction ratio is proportional to the square root of the product of the series inductance and shunt capacitance of a transmission line, the basic idea behind the design is to simultaneously increase the total inductance and capacitance values of the lumped model in Fig. 2.1(b). Meanwhile, to improve the miniaturization capability while not perturbing the line impedance, the ratio of the two terms (series inductance and shunt capacitance), representing the characteristic impedance of the synthesized line, should be kept as a specific value in accordance with (1.18).

In a very rough approximation, one can apply the principle of node exchange to simplify the design procedure by merging all series inductances and shunt capacitances as an L-section to approximately calculate the characteristic impedance (Z_c) and guided wavenumber (β_g) of the synthesized line as

$$Z_c = \sqrt{L_{tot}/C_{tot}} \tag{2.1}$$

$$\beta_g = \omega\sqrt{L_{tot}C_{tot}}/l. \tag{2.2}$$

l is the physical length of the line, and L_{tot} and C_{tot} are the direct sum of the series inductances and shunt capacitances as

$$L_{tot} = L_1 + L_2 + L_3 \tag{2.3}$$

$$\begin{aligned} C_{tot} &= C_{l1} + C_{l2} + C_{l3} + C_{l4} + C_{p1} + C_{p2} \\ &\quad + C_{p3} + C_{p4} + C_{s1} + C_{s2} + C_{s3} + C_{s4}. \end{aligned} \tag{2.4}$$

As L_{tot} and C_{tot} in (2.3) and (2.4) rise proportionally, the guided wavenumber in (2.2) is increased accordingly, but the characteristic impedance in (2.1) remains unaltered. It, in turn, effectively shortens the guided wavelength λ_g and hence significantly reduces the required physical length of a line with a given characteristic impedance and electrical length (say, 50 Ω and 90°). One must keep in mind that although the L-section network provides intuitive insight on how a synthesized line works, it is a very rough approximation since the original symmetric network is equivalent by an asymmetric one. For better approximation we should start with a T- or π-network; the synthesis details will be introduced in the next section.

In the design phase, extracting and modeling the quasi-lumped elements is another important issue to be considered. To do so, an EM full-wave simulator is first applied to calculate the *S*-parameters of each individual element in Fig. 2.1(b). The *S*-parameters are then converted to the *ABCD* matrices as well as to their corresponding π - or T- equivalent circuits. The required equivalent inductance and capacitance values are therefore extracted from the π - or T-equivalent models. It is well known that a high-impedance line in meander or spiral shape is able to increase the inductance value, while the capacitance value can be easily controlled by either tuning the finger lengths of an interdigital capacitor or by trimming the occupied area of a parallel-plate capacitor.

Following the design rule and extraction procedure, a synthesized microstrip line with a characteristic impedance of 50 Ω and an electrical length of 90° is designed on a 20-mil RO4003C substrate as in Fig. 2.1(a). The operating frequency is 915 MHz. For the given specification, the required total inductance and capacitance values, L_{tot} and C_{tot}, are first solved uniquely by (2.1) and (2.2). The required inductance for each meander-line inductor is therefore determined. The interdigital capacitors are then fixed by setting the desired transmission zero frequency as

$$f_{TZ} = \frac{1}{2\pi\sqrt{2L_2C_1}}. \tag{2.5}$$

The parallel-plate capacitors are the final parameter to be dealt with so as to compensate for the difference between the targeted total capacitance value and those contributed by the parasitic effects of the line inductors and interdigital capacitors. If required, design curves of each quasi-lumped element could be plotted using the aforementioned extraction procedure. After integrating all quasi-lumped elements, a post-integration tuning process may be required to take into account the parasitic coupling between elements, as well as the rough approximate nature of the L-section model. For this specific design, the final inductance and capacitance values are summarized in Table 2.1 for easy reference. The overall size is $0.015\,\lambda_g \times 0.015\,\lambda_g$. From (1.21) and (2.2), the slow wave factor of the synthesized microstrip line are calculated as

$$SWF = \frac{\beta_g}{k_0} = 5.9. \tag{2.6}$$

The synthesized microstrip line is characterized by its complex propagation constant and characteristic impedance. In accordance with (1.44) and (1.45), both parameters are extracted using the simulated/measured two-port *S*-parameters of the line. Referring to [1], the simulated and measured characteristic impedances at 915 MHz are 50.6 Ω and 49.7 Ω, respectively. Within a bandwidth of 530 MHz, the variation of the line impedance is less than ±5%. The measured phase delay is −89.7° at 915 MHz, and the slow wave factor is 5.8. Over the frequency band of interest, the variations of the normalized attenuation constant (α_g/k_0) and slow wave factor (β_g/k_0) are less than 5% and 0.03%, respectively. The phase response is also linear. Table 2.2 summarizes the calculated, simulated, and measured characteristic impedances and

Table 2.1 Lumped values of the synthesized microstrip line in Fig. 2.1

Element	Value	Element	Value	Element	Value
L_1-L_3	3.9 nH	C_{I1}, C_{I4}	0.26 pF	C_{p1}, C_{p4}	0.21 pF
C_1	0.22 pF	C_{I2}, C_{I3}	0.52 pF	C_{p2}, C_{p3}	0.36 pF
				C_{s1} - C_{s4}	0.51 pF

Table 2.2 Calculated, simulated, and measured characteristic impedances and slow wave factors of the 50-Ω synthesized microstrip line

	Calculation	Simulation	Measurement
Z_c	49.7	50.6	49.7
SWF	5.88	5.83	5.76

slow wave factors. The calculated ones are based on the lumped model using (2.1) and (2.6). Good agreement between the calculated, simulated, and measured results is observed.

Using a similar design approach, synthesized microstrip lines with a variety of characteristic impedances, say, 35 and 70 Ω, can also be developed. As a rule of thumb, the slow wave factors of the 35-, 50-, and 70-Ω lines, in the design examples in [1], are 4.6, 5.8, and 5.1. The values are remarkably larger than that of a conventional microstrip line, which is 1.7 over the frequency band of interest. The losses per guided wavelength, on the other hand, are found to be 0.85, 0.95, and 0.9 dB/λ_g for the 35-, 50-, and 70-Ω synthesized microstrip lines, respectively. They are inevitably higher than that of a conventional microstrip line, which is 0.23 dB/λ_g. The higher transmission loss of this non-periodic synthesized line is attributed to the large number of discontinuities used in the design.

2.2.2 90° and 180° Hybrid Couplers

A 3-dB quadrature hybrid, or a branch-line coupler, consists of pairs of 50-Ω and 35-Ω 90° lines. By replacing the conventional lines with synthesized microstrip lines developed in Sec. 2.2.1, a miniaturized quadrature hybrid is realized, as shown in Fig. 2.2. The simulated and measured *S*-parameters, which are found to be in good agreement, are illustrated in Fig. 2.3. The measured transmission coefficients ($|S_{21}|$, $|S_{31}|$) at the center frequency are −3.42 and −3.72 dB. The phase imbalance is less than ±0.2° from 880 to 935 MHz. The dimension of the coupler is 0.12 λ_g × 0.13 λ_g (λ_g: the guided wavelength of a conventional 50-Ω microstrip line at the same frequency). It is 27% the size of its conventional counterpart. Figure 2.4 compares the wideband response of this quadrature hybrid to that of a conventional design from 0.5 to 12 GHz. Benefitting from the allocated transmission zero as well as additional cross coupling effects, for this quasi-lumped design the bandwidth with signal attenuation better than 15 dB is from 2.2 to 11.1 GHz, or more than two octaves. This feature is not feasible by using conventional uniform transmission lines.

The rat-race coupler, or the 180° coupler, consists of six 70-Ω 90° transmission lines. Following the same design rule, a miniaturized rat-race coupler operated at 915 MHz is

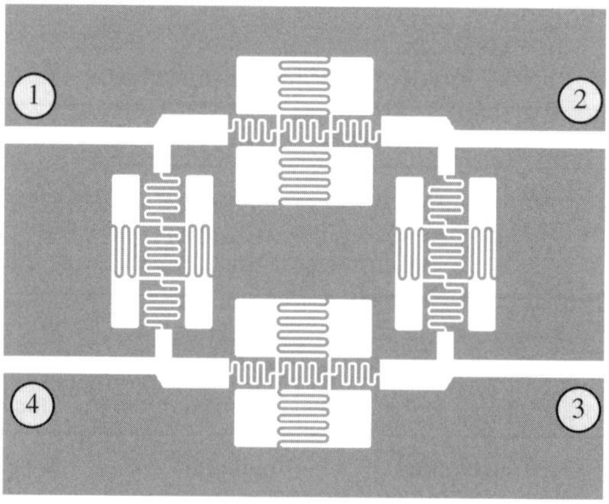

Figure 2.2 Photograph of the miniaturized quadrature hybrid. Source: Wang 2007 [1]. Reproduced with permission of IEEE

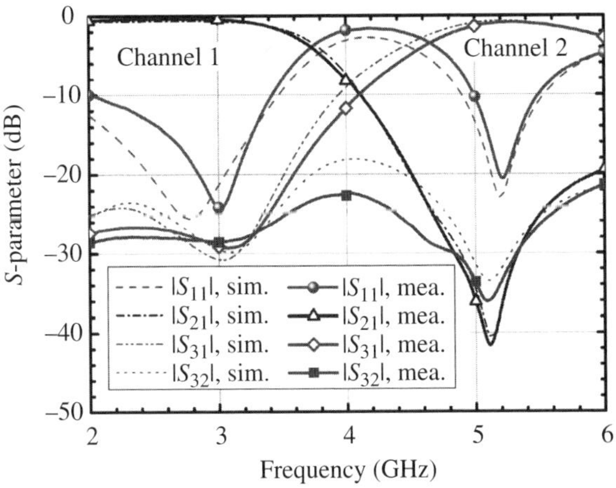

Figure 3.24 *S*-parameters of the diplexer on PCB process

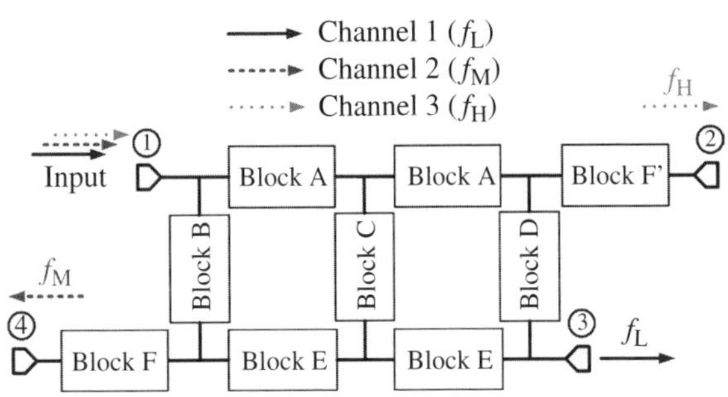

Figure 3.25 Schematic diagram of the triplexer

building blocks is reversed in the high band (f_H), in which the Block 1 becomes open-circuited whereas the Block 2 sets up the transmission channel (Channel 2). Benefiting from the dual operational states of both blocks, the design scheme requires no additional matching unit, hence simplifying the circuit complexity with reduced power dissipation. With a direct integration of the dual-mode right-handed and left-handed synthesized lines in Section 3.3.2, the circuit layout of the diplexer is shown in Fig 3.23(b).

Figure 3.24 shows the *S*-parameters of the diplexer. The in-band transmission loss is less than 1 dB, and the isolation between the outputs is 30–40 dB. The input-to-inactive channel suppression is in the same order. With the criteria of $|S_{11}|<-10$ dB, ($|S_{21}|$, $|S_{31}|$)>-1 dB (in-band) and ($|S_{21}|$, $|S_{31}|$, $|S_{32}|$)<-25 dB (out of band), the fractional bandwidth is 53.8% in the Channel 1 (2.01–3.3 GHz) and 6.7% in the Channel 2 (5.1–5.45 GHz).

3.5.2 Triplexer

In this subsection, the dual-/tri-operational mode synthesized lines in Secs. 3.3 and 3.4 are adopted to realize a triplexer with a very unique topology. Figure 3.25 shows the schematic diagram of the triplexer as well as the signal flows in the operating bands. The triplexer

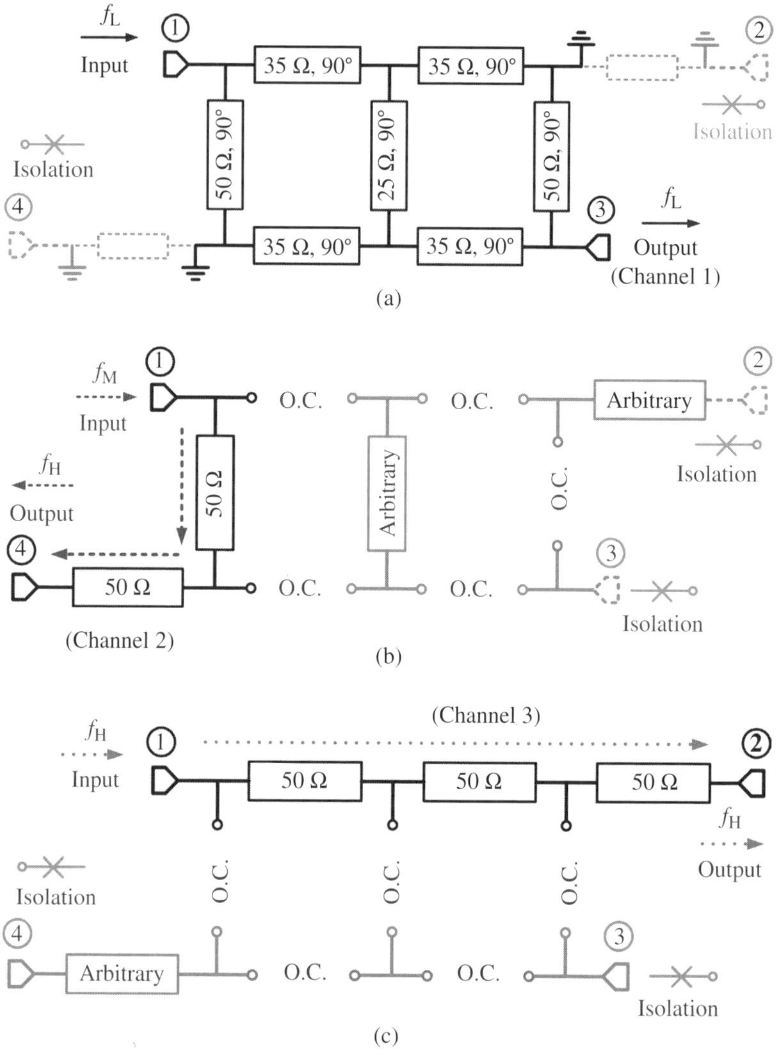

Figure 3.26 Operational principles of the triplexer in the (a) low band (f_L), (b) mid band (f_M), and (c) high band (f_H)

consists of six building blocks, each associated with a unique circuit response in the three bands. For simplicity, the building blocks are termed as the Blocks A, B, C, D, E, F, and F'. Here, the Block F' is a variation of the Block F with the same circuit topology.

The operational principle is explained in detail in Fig. 3.26(a)–(c). The required electrical responses of the building blocks are shown at the same time. The design basis is a planar crossover using two quadrature hybrids in cascade. Each line section in the crossover is replaced by a tri-mode synthesized transmission line, and two extra synthesized units are added to the terminals 2 and 4. In the design, the common input is the port 1, and the remaining ports (ports 2, 3, and 4) are denoted as the Channels 1, 2, and 3 of the triplexer.

4

Applications to Heterogeneous Integrated Phased Arrays

C. H. Lai and T. G. Ma

4.1 Introduction

Phased arrays play a crucial role in modern wireless communication systems due to their fantastic abilities to shape, switch, and/or steer the radiation beam of antennas. Without using digital signal processing units, beam-switching, beam-steering, and even multi-beam arrays can be realized at microwave and millimeter-wave frequencies using RF circuit components. As examples, the Butler matrix [1–8], the Blass matrix [9], and the Nolen matrix [10] are the common ways to realize a beam-forming network with multiple selectable beams. Among which, the Butler matrix, named after its inventor, is especially of interest due to its low-loss nature, cost efficiency, and compactness.

The retrodirective array (RDA) presents a dramatically different scenario [11–20]. It has the ability to retransmit the interrogation signal back to the interrogator without any *a priori* knowledge of the direction of arrival. For this purpose, the operation of an RDA requires the phase difference between the retransmitted signals at adjacent antenna terminals to be the negative of the phase difference between the incident ones. This is called the *phase difference reversal condition*. The Van Atta arrays [11–14] and the phase-conjugating arrays (PCAs) [15–19] are the most common approaches to fulfill the requirements. The Van Atta array connects the array elements into pairs with the phase adjustment lines inserted in-between. The length of each adjustment line must be an integer multiple of the guided wavelength to ensure the phase difference reversal condition. The phase-conjugating array (PCA), on the other hand, adopts heterodyne conversion with mixers and local oscillators for phase reversal at the terminals. In [20], the phase adjustment lines were replaced by quadrature hybrids, crossovers, and reflective loads, hence giving rise to a linear retrodirective array termed as *reflection-type retrodirective array*. In essence, it is a variation of the Van Atta array.

Synthesized Transmission Lines: Design, Circuit Implementation, and Phased Array Applications, First Edition.
Tzyh-Ghuang Ma, Chao-Wei Wang, Chi-Hui Lai and Ying-Cheng Tseng.

A phased array generally requires a complicated feeding network whose architecture is highly depended on the application purpose. Due to the uniqueness of each of the aforementioned feeding topologies, in traditional thinking it would be a very challenging task if one would like to integrate multiple phased arrays into a single network without dramatically increasing the overall size and fabrication cost. Nevertheless, it is recently noticed that, with the multi-operational mode synthesized transmission lines introduced in Chapter 3, the ambition to integrate multiple arrays as a single system has come true.

In this chapter, several innovative heterogeneous integrated phased array systems will be introduced. It involves beam-switching array, Van Atta array, reflection-type retrodirective array, and phase-conjugating array. First of all, dual-mode retrodirective arrays, enabling dual-band operation for increasing the data throughput and high quality of service, are introduced. What follows is a heterogeneous integrated phased array functioning as a Butler matrix and a reflection-type retrodirective array respectively in two different bands. A tri-mode heterogeneous integrated phased array, serving as a beam-switching array, a Van Atta array, and a phase-conjugating array respectively in the low, mid, and high bands, is then implemented and demonstrated. A simplified version using two antenna elements is briefly discussed at the end of the chapter.

4.2 Dual-Mode Retrodirective Array

4.2.1 Design Goal

The first example is a dual-mode (dual-band) retrodirective array. The demand of high-speed data transmission increases continuously in modern communication systems. Dual-band operation, enabling data transmission in two individual frequency bands, provides a simple yet efficient approach to increase the data throughput. An interrogation system can dramatically improve its performance by dual-band operation, as well. A retrodirective array (RDA) capable of dual-band operation would be, therefore, of great benefit. Figure 4.1 depicts the design goal of the heterogeneous integrated array presented in this section. It is a retrodirective array capable of dual-band operation with two different architectures but sharing the same physical configuration [21]. Both bands can be used for data communication; or alternatively, one allocated for data transmission and the other one for wireless power transmission is also a possibility.

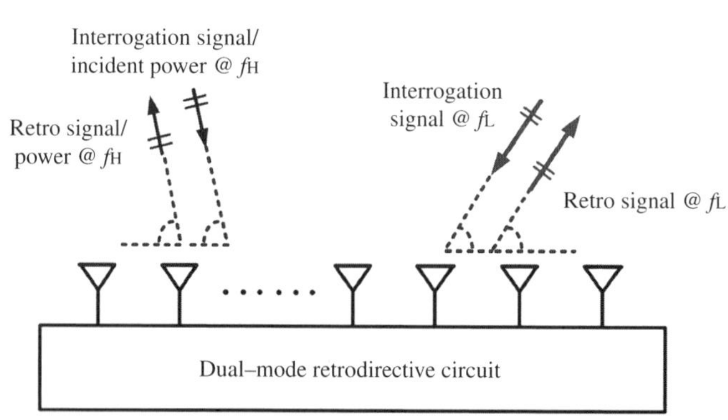

Figure 4.1 Design goal of a dual-mode retrodirective array

4.2.2 System Architecture

The architecture of the dual-mode retrodirective array is illustrated in Fig. 4.2(a). The design is fulfilled by two unusual four-port networks: the dual-mode quadrature coupler and dual-mode crossover. Their schematic diagrams are shown in Fig. 4.2(b). Referring to the figure, the dual-mode retrodirective array can be divided into four parts. The largest portion comprises the dual-mode couplers and crossovers, arranged similar to a reflection-type retrodirective array. The remaining parts include the matching sections, open stubs, and phase adjustment lines.

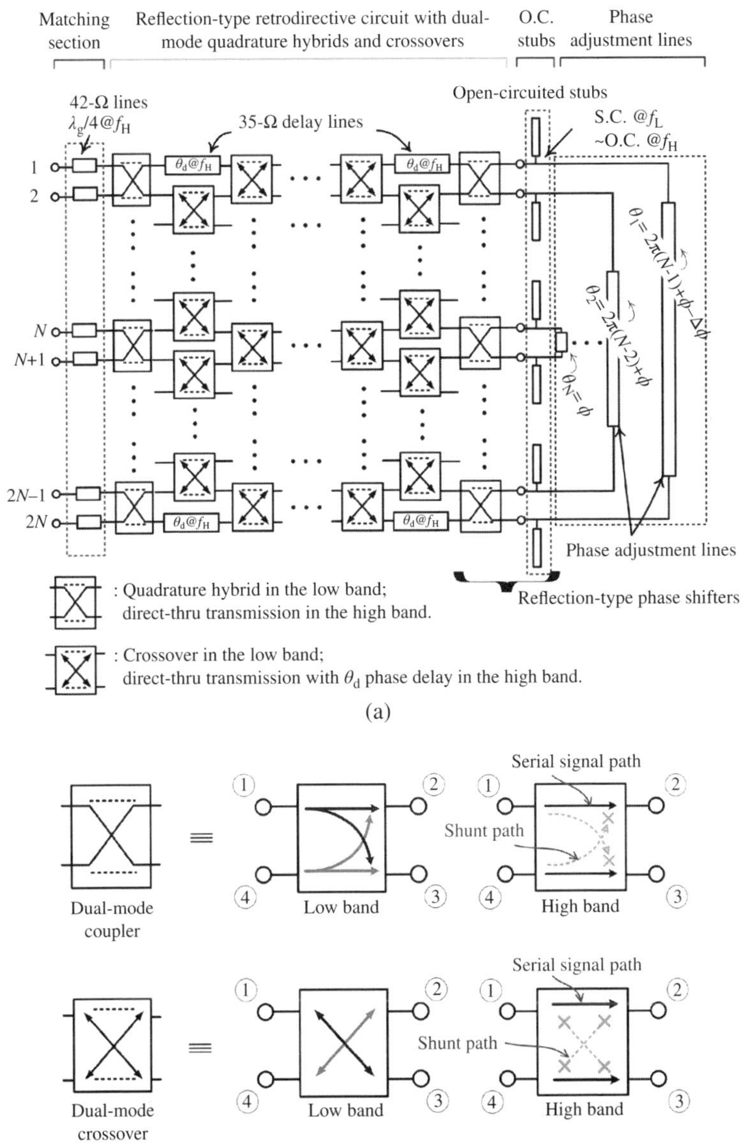

Figure 4.2 (a) Schematic diagrams of the dual-mode retrodirective array and (b) the core components – dual-mode coupler and crossover. Source: Tsai 2011 [21]. Reproduced with permission of IEEE

The dual-mode coupler is operated similar to a conventional quadrature hybrid in the low band (f_L); it, however, becomes identical to two isolated direct-thru transmission lines in the high band (f_H) [22]. The dual-mode crossover is a cascade connection of two dual-mode couplers with additional connecting lines in-between. Similar to the coupler, the dual-mode crossover functions identical to its conventional counterpart in the low band, with the input signal at port 1 completely transferred to port 3. On the other side, in the high band, the shunt paths are open-circuited such that the network is again identical to a pair of uncoupled thru transmission lines.

Referring again to Fig. 4.2(a), the rightmost quadrature hybrid of the network, together with the open stubs, functions as a reflection-type phase shifter in the low band. The open stubs are a quarter wavelength at f_L to provide the required total reflection at the end terminals of the couplers ($\Gamma=-1$). Since the open stubs are nearly a half-wavelength long in the high band, they are invisible to the high-frequency signals. The outputs of the rightmost quadrature hybrids, in the high band, are therefore connected as pairs by the phase adjustment lines of the retrodirective circuit. This enables the Van Atta array configuration. The final part of the retrodirective array is the matching section on the left side. As indicated in Fig. 4.2(a), they are no more than common quarter-wavelength transformers.

Figure 4.3(a) and (b) further explains the operation principles of the design using signal flows. In Fig. 4.3(a), the dual-mode couplers and crossovers behave exactly the same as their conventional counterparts in the low band. At the center frequency, the $\lambda_g/4$ open stubs are shorted out at the end terminals such that the hybrids at the final stage turn into reflection-type phase shifters. Since the interrogation signal injected into any input port will pass through the delay line once and only once during the retransmission process, the exact electrical lengths of the connecting lines in the first and 2*N*th paths are not of concern. The characteristic impedance of the lines, however, should be 35 Ω to guarantee the impedance matching.

Meanwhile in the high band, all shunt signal paths in the dual-mode couplers and crossovers are prohibited, making the retrodirective array equivalent to a Van Atta array as shown in Fig. 4.3(b). The delay lines in the outermost paths (1st and 2*N*th paths) are used to equalize the phase delay of all signal paths in the network. The electrical length of the delay line, θ_d, is equal to the phase delay of a dual-mode crossover in the direct-thru mode at f_H. The 42-Ω matching sections at the input terminals serve as the quarter-wavelength transformers for the 50-Ω system impedance.

4.2.3 *Circuit Realization*

A four-element dual-mode retrodirective array is realized as a demonstration. The circuit layout is shown in Fig. 4.4. The center frequencies of the low and high bands are 2.05 and 3.95 GHz. The key components, dual-mode couplers and crossovers, are developed using the synthesized transmission lines in Sec. 3.3.4 to blockade the shunt signal paths in the high band. By following the same synthesis procedures in Chapter 2, the 35-Ω lines in the serial paths are also replaced by synthesized transmission lines for size reduction.

The folding method in [20] is applied to illustrate the electrical response of the network. A schematic diagram of the unfolded retrodirective circuit is shown in Fig. 4.5(a). The unfolded circuit is realized by first mirroring the retrodirective circuit with respect to the vertical mirror plane, followed by shifting the quadrature hybrids at the last stage of the circuit to the first one. The unfolded retrodirective circuit becomes an eight-port network, with each section composed

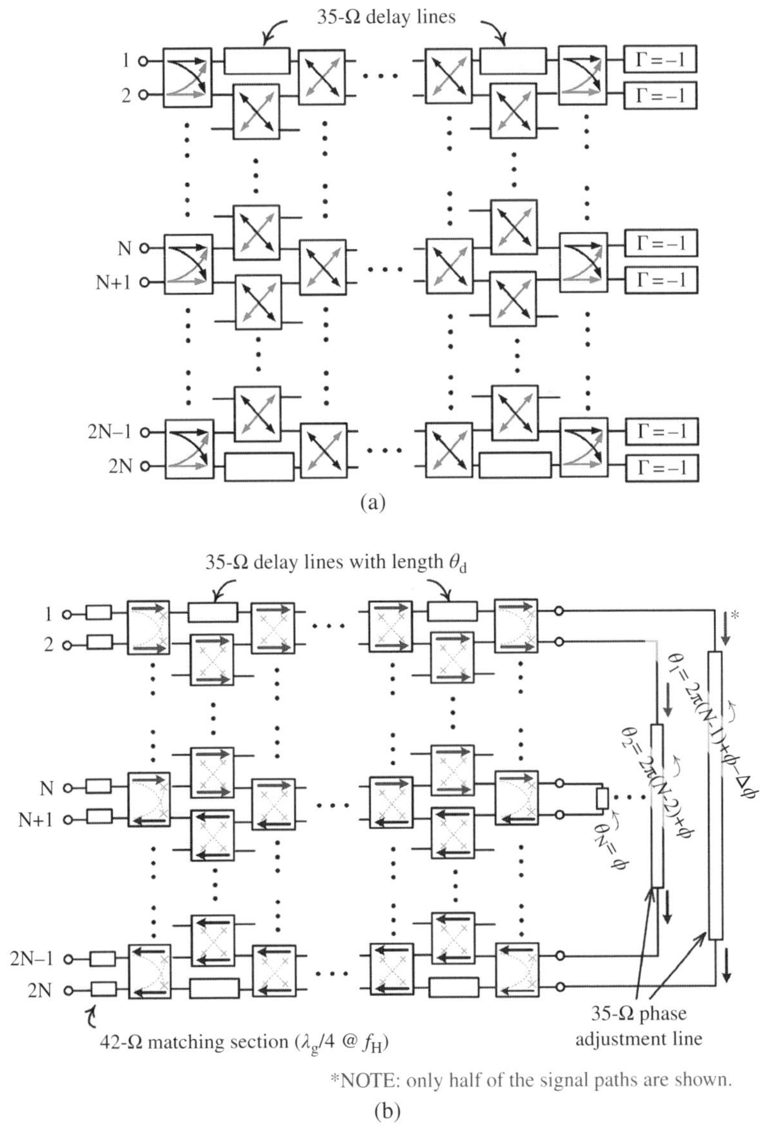

Figure 4.3 Operational principle of the dual-mode retrodirective array in the (a) low and (b) high bands. Source: Tsai 2011 [21]. Reproduced with permission of IEEE

of either two crossovers, or a crossover with two delay lines. The phase adjustment lines and the open-circuited stubs are removed from the unfolded circuit to simplify the analysis. This removal is trivial since in the high band the open stubs are invisible and the phase adjustment lines simply introduce extra but equal delays to the retransmitted signals. In the low band, meanwhile, the open stubs make the phase adjustment lines become invisible in the mirror plane.

Figure 4.5(b) illustrates the signal flows of the unfolded four-element retrodirective circuit operated in the low and high bands. Each crossover is represented by a pair of crossed signals. In the low band, the signal injected into port 1 is delivered to port 8; the signal at port 2 is

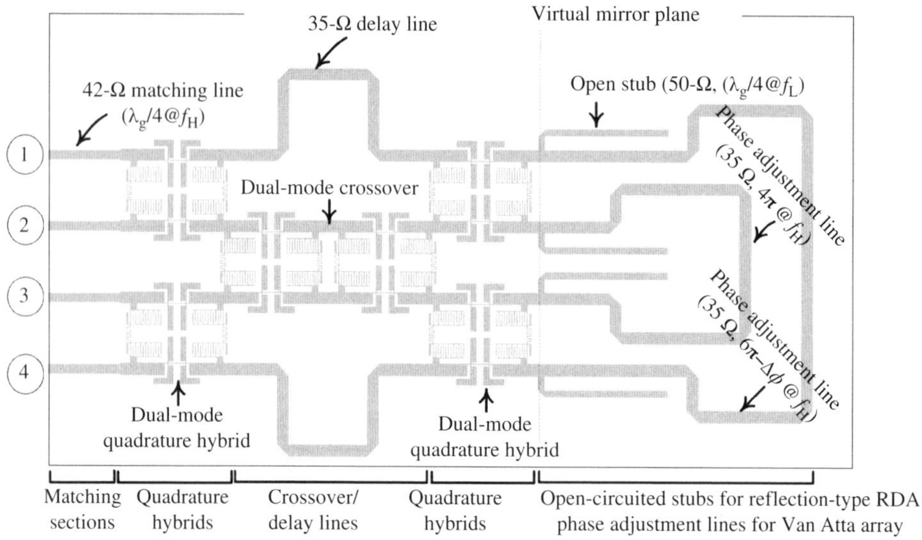

Figure 4.4 Circuit layout of a four-element dual-mode retrodirective array

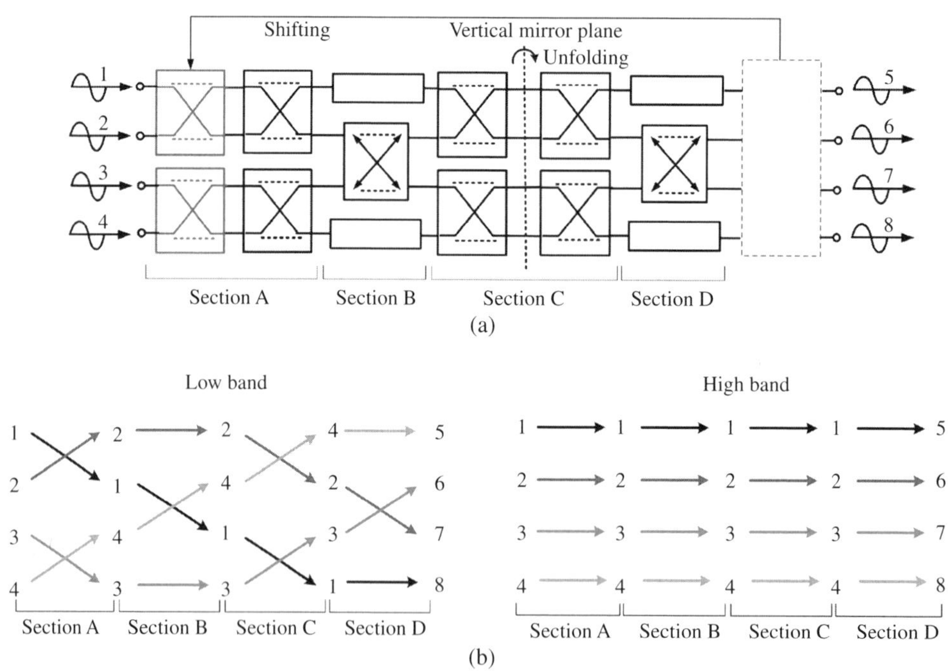

Figure 4.5 (a) Schematic diagram and (b) signal flows of the unfolded four-element dual-mode retrodirective circuit in the low and high bands. Source: Tsai 2011 [21]. Reproduced with permission of IEEE

transmitted to port 7, and so forth. It fulfills the phase difference reversal condition required for a retrodirective array.

In the high band, since the crossed paths are blocked away from the serial transmission paths, the input signal at port 1 is delivered directly to port 5, and so on. By incorporating with the phase adjustment lines, in a real folded four-element dual-mode retrodirective array, the input signal at port 1 is actually redirected to port 4 as the retransmitted signal. It therefore fulfills the phase difference reversal condition for a Van Atta array.

With a realization of the unfolded dual-mode retrodirective circuit, the eight-port S-parameters are measured and illustrated in Fig. 4.6. In the low band, the transmission coefficient in each of the retransmitting path ($|S_{81}|$, $|S_{72}|$, $|S_{63}|$, and $|S_{54}|$) is almost identical; it is −2.4 dB at the center frequency. The isolations are better than 21 dB. The phase difference between the inner and outer transmission paths ($\angle S_{81} - \angle S_{72}$) is almost zero.

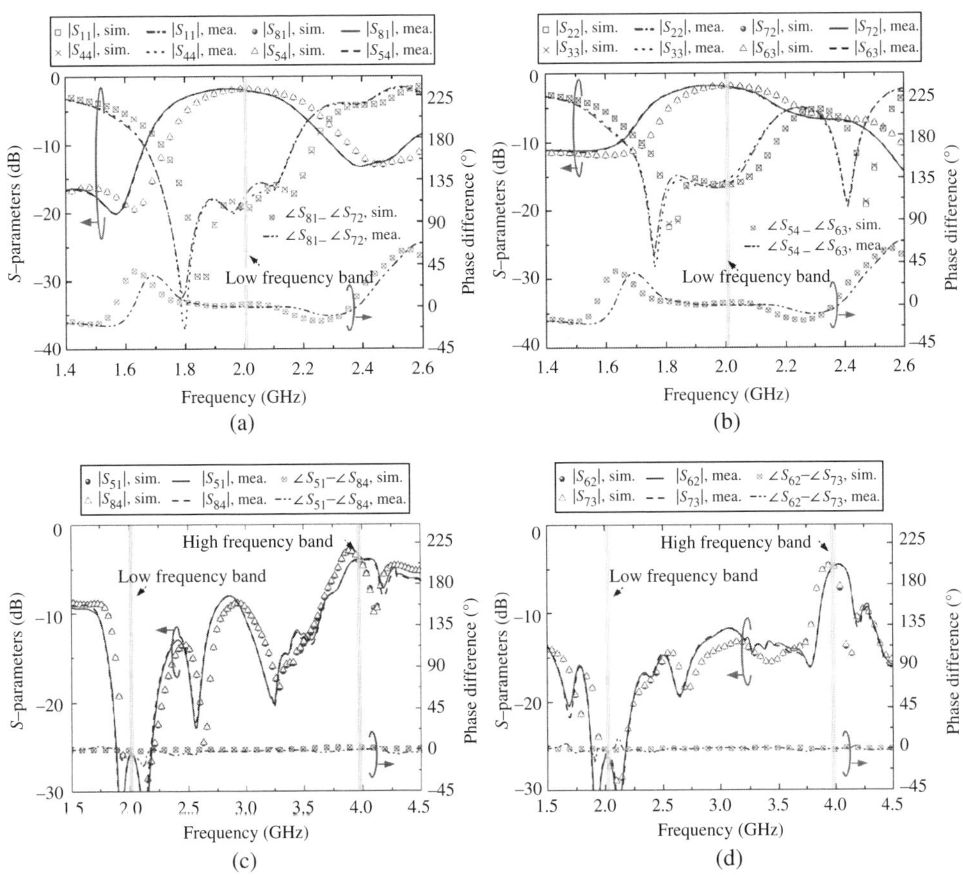

Figure 4.6 S-parameters of the unfolded four-element dual-mode retrodirective circuit: (a) ports 1 and 4 in the low band; (b) ports 2 and 3 in the low band; (c) ports 1 and 4 in the high band; and (d) ports 2 and 3 in the high band. Source: Tsai 2011 [21]. Reproduced with permission of IEEE

In the high band, however, the transmission loss in the outermost paths ($|S_{51}|$ and $|S_{84}|$) is lower than that in the inner paths ($|S_{62}|$ and $|S_{73}|$). At 3.95 GHz, it is −4.0 dB in the outer path, in contrast to −5.3 dB in the inner one. The discrepancy is due to the fact that the two paths are not identical. The number of synthesized lines in the inner path is twice of that in the outer one. The effects of parasitic loss of series elements and the non-ideal reactive loading of shunt elements (not an ideal open) are doubled, and hence make the differentiation. The isolation between paths is at least 15 dB, and can be up to 50 dB if they are not adjacent to each other. The phase difference between the inner and outer paths ($|\angle S_{51} - \angle S_{62}|$) is around 15°. The small deviation from the ideal value (0°) can be compensated for by trimming the outermost phase adjustment line, as illustrated in Fig. 4.4 with the compensation term $\Delta\phi$.

4.2.4 Bistatic Radiation Patterns

The retransmitted (or bistatic) radiation patterns of the dual-mode retrodirective array are experimentally verified using the experimental setup in Fig. 4.7. A photograph of the complete dual-mode array is shown in Fig. 4.8. Apart from the dual-mode feeding network, this array is composed of gain amplifiers, circulators, DC bias boards, equal-phase connection lines, and transmitting/receiving patch arrays. The details of the auxiliary components in the system can be found in [21].

Referring to Fig. 4.7, the measurement is arranged in a bistatic configuration. The illuminating antenna is placed on a horizontal foam rod to transmit the interrogation signal in several prescribed directions. The distance between the illuminating antenna and the retrodirective array satisfies the far-field condition. To reduce the nearby coupling and interference, the electric field polarizations in the illuminating and retransmitting paths are orthogonal to each other to improve measurement accuracy.

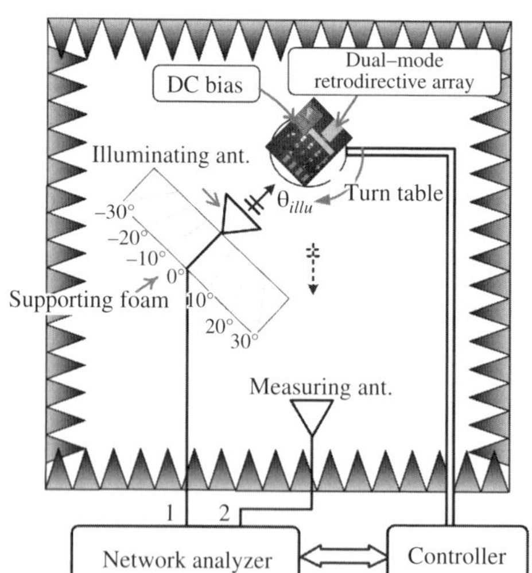

Figure 4.7 Schematic diagram of the experimental setup for measuring the bistatic radiation patterns of the dual-mode RDA

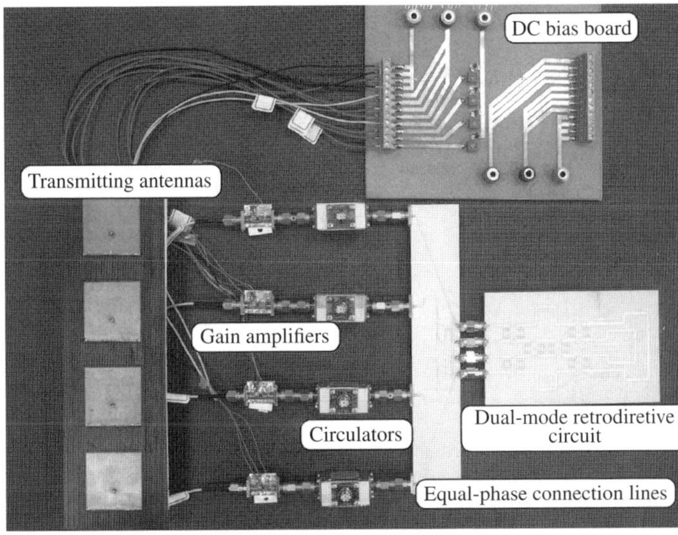

Figure 4.8 Photograph of the four-element dual-mode retrodirective array including the bias network, gain stage, and patch antennas. Source: Tsai 2011 [21]. Reproduced with permission of IEEE

The measured bistatic radiation patterns in the low and high bands are illustrated, respectively, in Fig. 4.9(a) and (b). The illuminating direction (θ_{illu}), indicated by an arrow in each of the figures, varies from −20° to +20° with a 10° step. The retransmitted signal is directed back to the direction of arrival (i.e., illuminating antenna) in both low and high bands as expected. The measured sidelobe levels, however, are considerably larger than those in the calculation. The discrepancy can be alleviated by calibrating out the noise from the cluster scattering due to nearby objects. The results will be shown in the following subsection, Sec. 4.2.5.

4.2.5 Alternative Architecture

By replacing the phase adjustment lines in Fig. 4.4 with drain mixers [23], an alternative four-element dual-mode retrodirective array is realized. It is an integration of the reflection-type RDA and the PCA. The circuit layout is shown in Fig. 4.10(a). The low-band operation is the same as the previous case. In the high band, since the crossover paths are also blocked away due to the unique properties of the couplers and crossovers, the input signals at ports 1–4 are fed directly to the drain mixers. After being mixed with a local oscillator, the IF signals with reversed phase response will be retransmitted to ports 1–4 as the output signals via the same serial paths. The signal flow is illustrated in Fig. 4.10(b).

The calculated and measured bistatic radiation patterns in the high band are shown in Fig. 4.10(c). The illuminating direction (θ_{illu}) varies from −30° to +30° with a 20° step. The retransmitted signal is again directed backward to the direction of arrival as the theoretical prediction. The agreement between the results is improved since the clutter noise has been calibrated out accordingly.

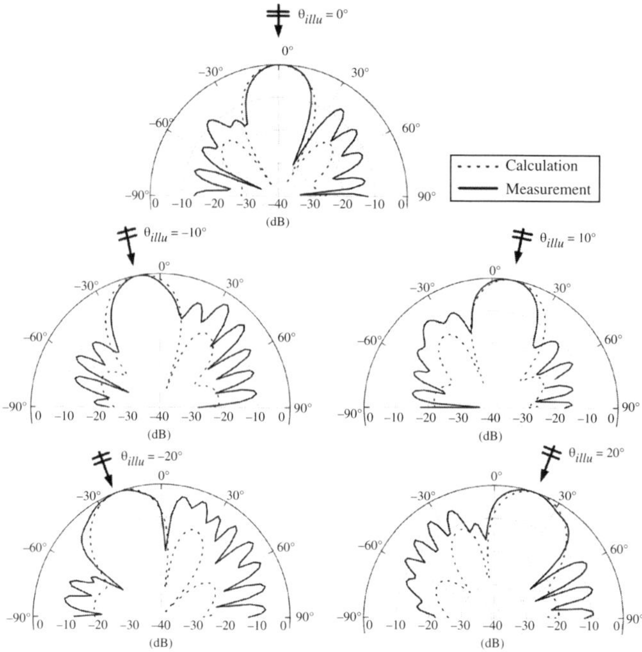

(a)

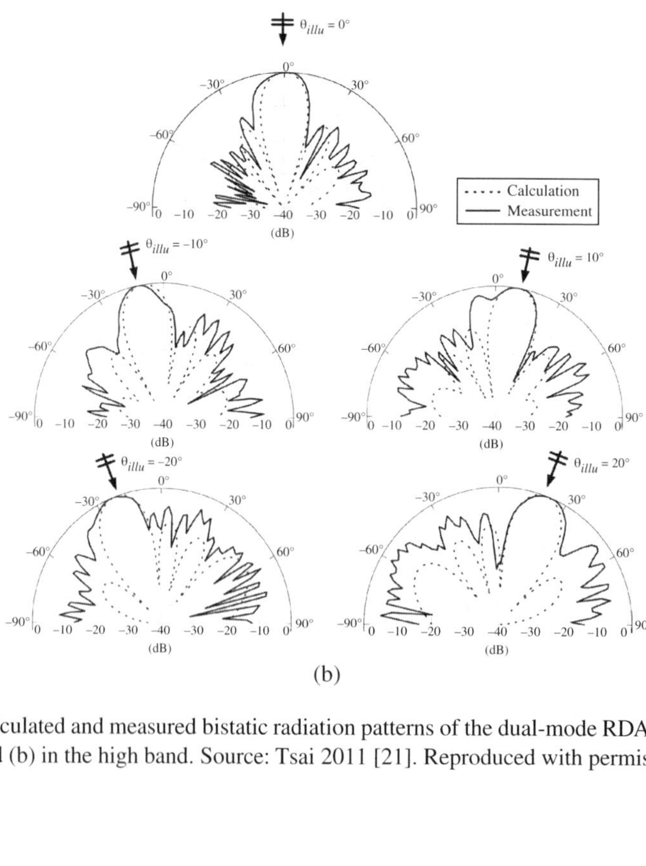

(b)

Figure 4.9 Calculated and measured bistatic radiation patterns of the dual-mode RDA in Fig. 4.4: (a) in the low band and (b) in the high band. Source: Tsai 2011 [21]. Reproduced with permission of IEEE

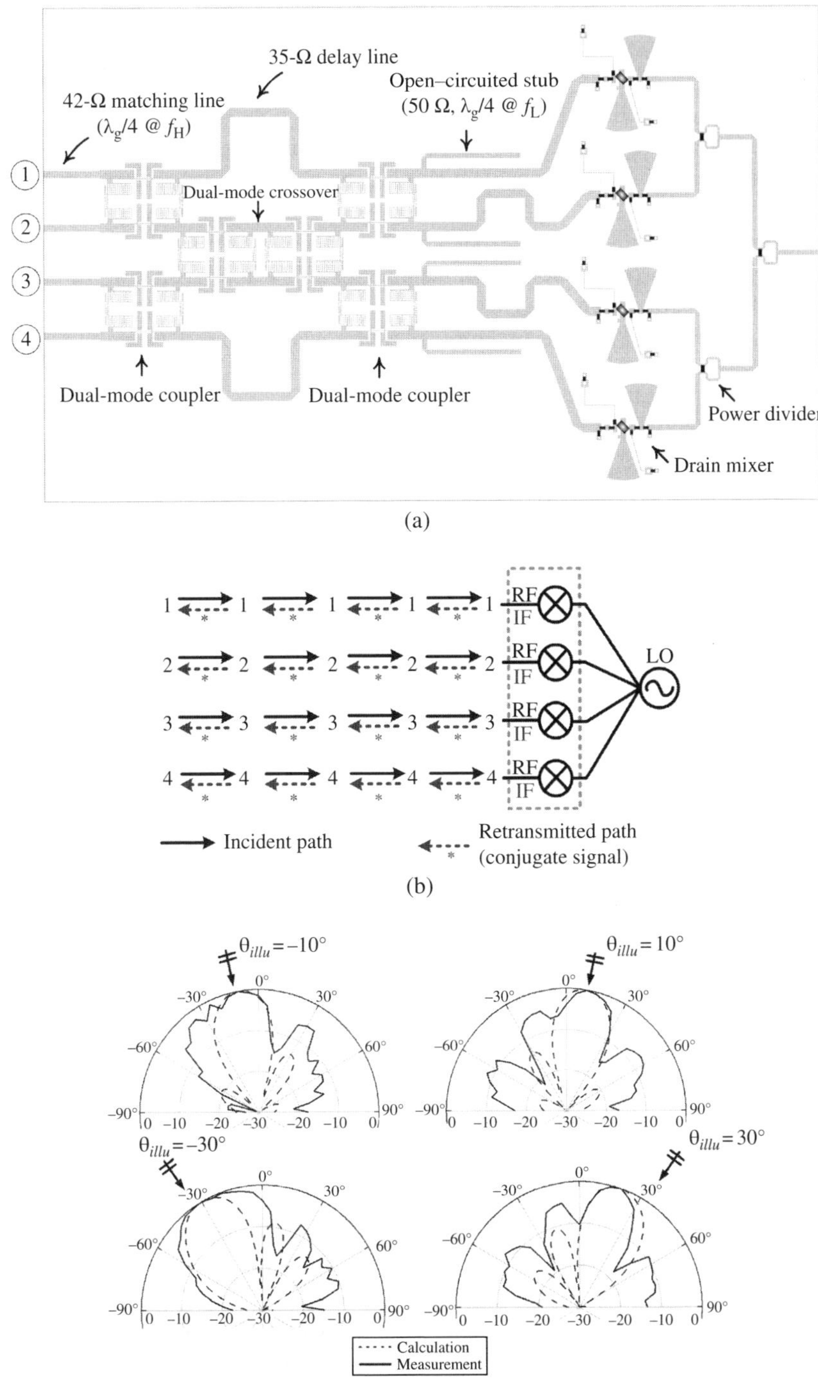

Figure 4.10 Four-element dual-mode retrodirective array using an alternative configuration with PCA: (a) circuit layout; (b) signal flows; (c) bistatic radiation patterns in the high band

4.3 Dual-Mode Integrated Beam-Switching/Retrodirective Array

4.3.1 Design Goal

Beam-switching and retrodirective arrays are unique and both play crucial roles in modern wireless communication systems. Referring to Fig. 4.11, imagine the possibility that a dual-mode integrated phased array capable of combining the functionalities of both arrays could be realized. In one mode, the system serves as a beam-switching array with duplex channels for communication and positioning. In the other mode, it becomes a retrodirective array capable of either acting as a passive transponder for responding information with low power consumption, or providing an efficient link for wireless power transmission. In this section, the design aspects for realizing such an interesting topic will be covered and discussed.

4.3.2 System Architecture

Figure 4.12 shows the schematic diagram of a four-element heterogeneous integrated array to fulfill the design goal [24]. The core component is the dual-mode feeding network. When the phased array is operated in the beam-switching mode, the feeding network is identical to a typical 4×4 Butler matrix, with the input part, consisting of a gain amplifier and an SP4T

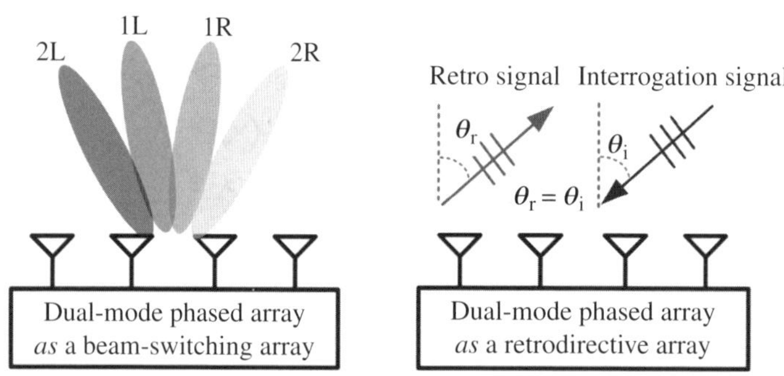

Figure 4.11 Design goal of the dual-mode heterogeneous integrated beam-switching/retrodirective array. Source: Zou 2013 [24]. Reproduced with permission of IEEE

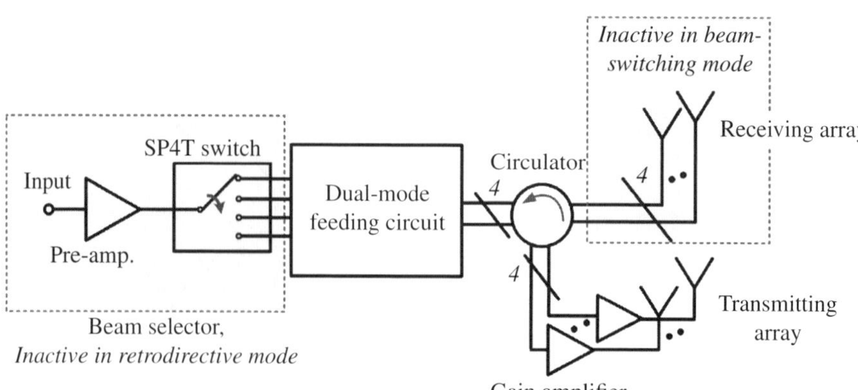

Figure 4.12 Schematic diagram of the four-element heterogeneous integrated beam-switching/RDA

switch, serving as the beam selector. The outputs of the feeding network are connected to the transmitting array via the circulators, while the receiving array is inactive.

When the array is operated in the RDA mode, the incoming signals from the interrogator are captured by the receiving array and fed to the dual-mode feeding network via the circulators. The dual-mode feeding network now functions as a reflection-type RDA such that the reentrant signals into the circulators satisfy the phase difference reversal condition and reradiate toward to the direction of arrival by the transmitting array.

The dual-mode feeding network in Fig. 4.12 is unique, and needs to be designed with innovative thought such that the topology can be automatically switched between two modes (frequencies) without using any active switching circuit. To this end, let us consider first the topologies of a typical four-element Butler matrix and a four-element reflection-type RDA in Fig. 4.13. It is surprising to find that the feeding circuits of the two, despite of their dramatically different applications, are very similar. The differences are limited to the additional crossover at the outputs of the Butler matrix, and the reflective loads at the inputs of the RDA network. Accordingly, it would be fruitful if the two components could be replaced with dual-mode ones which simultaneously produce the respective responses in the two frequency bands.

With this basic principle in mind, Fig. 4.14 shows the complete schematic of the feeding network for realizing the dual-mode heterogeneous integrated beam-switching/retrodirective

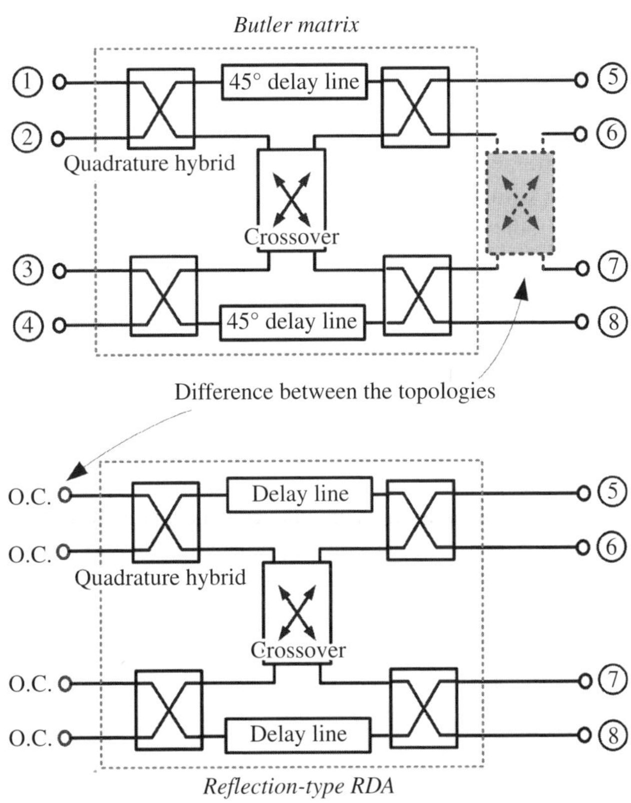

Figure 4.13 Comparison of the topologies of a typical four-element Butler matrix and a reflection-type RDA

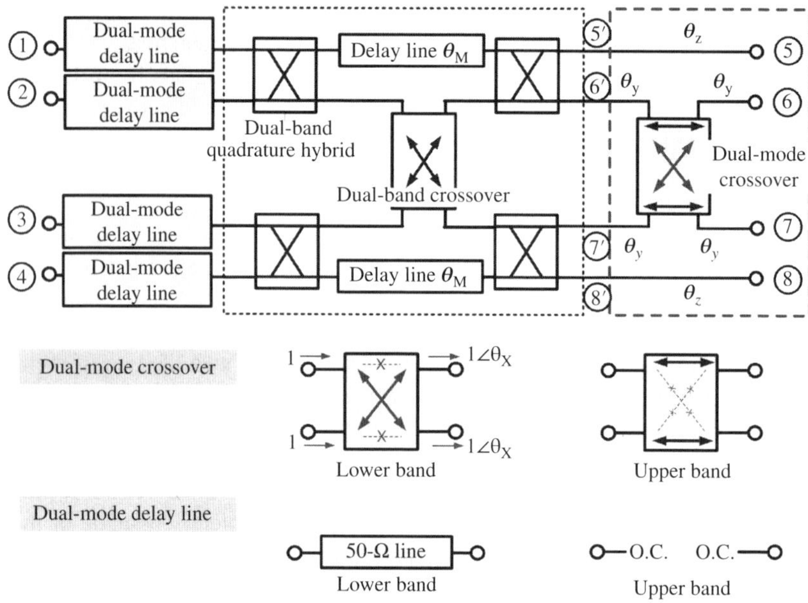

Figure 4.14 Schematic diagram of the dual-mode feeding network for the heterogeneous integrated beam-switching/retrodirective array

array. By comparing the network in Fig. 4.13 to that in Fig. 4.14, the common part, that is, the components in the dot frame, remains the same in both modes. They can be realized by a typical dual-band design with identical responses in the two bands. The dual-mode crossover in the dash part, on the other hand, functions as its conventional counterpart in the beam-switching mode, but behaves like a pair of uncoupled transmission lines in the RDA mode. In addition, the dual-mode delay lines at the inputs should serve as routine connecting lines in the beam-switching mode, but become perfectly open-circuited when looking into either side of the lines in the RDA mode. These specifications exactly coincide with the properties of the dual-mode synthesized transmission lines introduced in the first part of Chapter 3, making them the ideal building blocks for realizing the integrated phased array.

Nevertheless, it should be aware that for a phased array, the relative phase difference between adjacent output ports should always follow a specific pattern to guide the radiation beam to a desired direction. Introducing the dual-mode components inevitably destroys the phase relationship of the original feeding network. To tackle the problem, a phase compensation scheme by using extra connecting lines in the dual-mode crossover (θ_x) as well as between the hybrid couplers (θ_y, θ_z) is adopted, as shown in the dash part in Fig. 4.14. By using a simple calculation to retain the required phase relationships in both modes, the formulae for the connecting lines are given as

$$\theta_x = \left(\frac{1}{2} - \frac{n_2}{k} + 2n_1\right)\pi \tag{4.1}$$

$$\theta_y = \frac{\theta_z}{2} + \frac{2n_2 - k}{4k}\pi \tag{4.2}$$

$$n_1, n_2 = 0, 1, \cdots. \tag{4.3}$$

Here, θ_z is a degree of freedom that can be arbitrarily selected. The interested readers can find the detailed derivation of (4.1)–(4.3) in [24].

4.3.3 *Circuit Realization*

After determining the circuit specifications and required building blocks, a dual-mode feeding network for realizing the integrated beam-switching/retrodirective array is fabricated in Fig. 4.15. The operating frequencies are 2.4 and 5.15 GHz. The core components, including the dual-mode crossover and delay lines, are developed using the dual-mode synthesized lines with series LC resonators in Chapter 3. The dual-band hybrids, on the other hand, are realized by the port extension technique presented in [25].

The *S*-parameters in Fig. 4.16(a) clearly demonstrates the dual-mode operation capability of the crossover. At 2.4 GHz, the crossover path has a measured transmission coefficient ($|S_{31}|$) of –0.6 dB, while the reflection ($|S_{11}|$) and isolations ($|S_{21}|$, $|S_{41}|$) are lower than –18 dB. At 5.15 GHz, the measured transmission coefficient in the direct-thru path ($|S_{41}|$) becomes –0.5 dB, while the isolations ($|S_{21}|$, $|S_{31}|$) are better than 55 dB. The reflection coefficient ($|S_{11}|$) is –21.9 dB at this frequency.

The circuit responses of the dual-mode integrated feeding network as a Butler matrix are then depicted in Fig. 4.16(b). The port 1 is served as the excitation. At 2.4 GHz, the transmission coefficients are around –6.4 to –7.8 dB. The isolation is better than 18 dB. Similar results are observed when the port 2 is used as the input.

Figure 4.16(c) shows the *S*-parameters of the dual-mode feeding network in the RDA mode. At 5.15 GHz, the transmission coefficients along the path 5–8 and path 6–7 ($|S_{85}|$ and $|S_{76}|$)

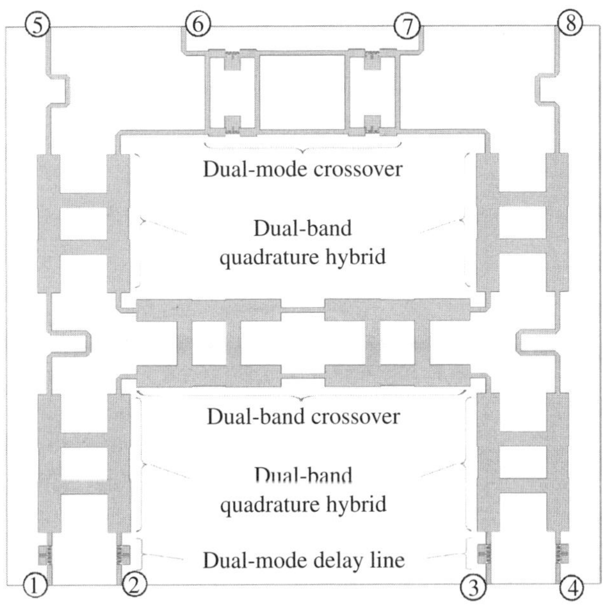

Figure 4.15 Circuit layout of the dual-mode feeding network for the four-element heterogeneous integrated phased array

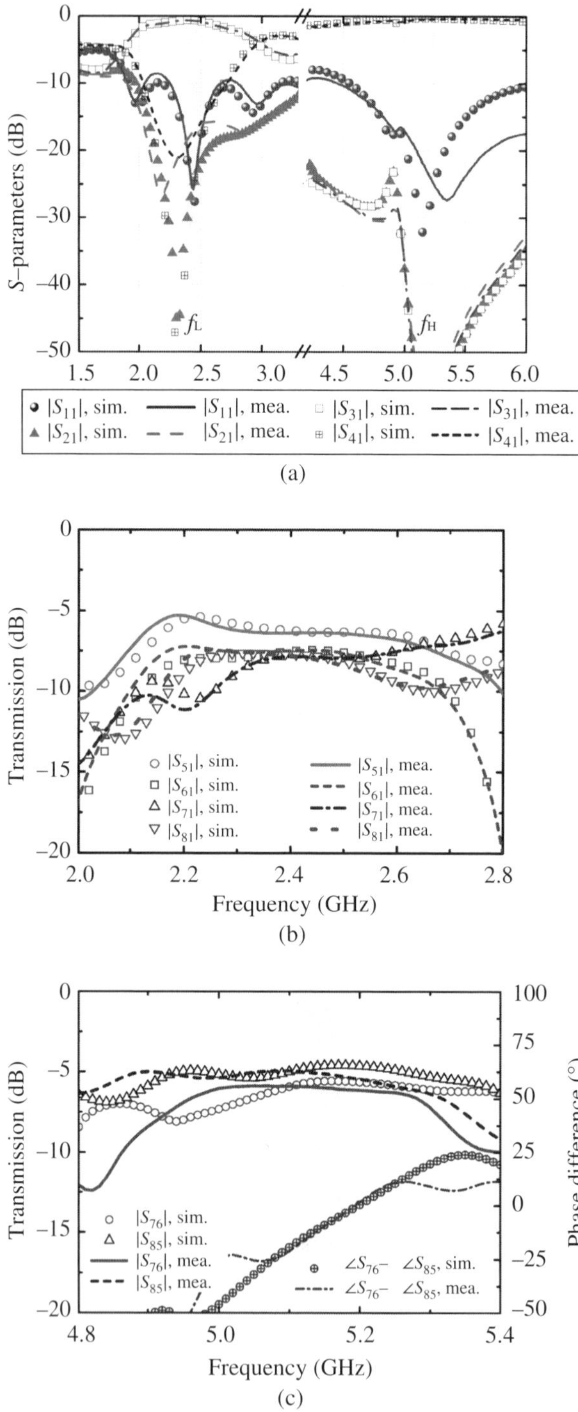

Figure 4.16 *S*-parameters of the (a) dual-mode crossover, (b) dual-mode feeding network in the beam-switching mode (port 1 excited), and (c) dual-mode feeding network in the RDA mode. Source: Zou 2013 [24]. Reproduced with permission of IEEE

are −6.0 and −5.1 dB, respectively. The slight difference in path loss can be accounted for by the parasitic reactive loading of the synthesized transmission lines (not a perfect open circuit). The isolations are better than 20 dB, and the phase difference between the paths is less than 10°.

4.3.4 *Radiation Characteristics*

Figure 4.17 depicts the radiation patterns of the heterogeneous integrated phased array in the beam-switching mode at 2.4 GHz. For the interested readers, a detailed description of the measuring setup as well as the auxiliary components used in the array can be found in [24]. In Fig. 4.17, depending upon the inputs, the main beam of the array is steered to 46°, 14°, −14°, and −44°, very close to the theoretical values. The measured peak gains are all around 12.5–13.5 dBi, making the proposed design suitable for positioning and location tracking.

The retransmitted radiation patterns in the RDA mode are measured bistatically at 5.15 GHz. The results are summarized in Fig. 4.18. The incident angle of the illuminating wave varies from −20° to +20° with a 10° step. The calculated data are achieved by first using the simulated *S*-parameters of the dual-mode circuit to compute the array factors. The array factors are then multiplied by the simulated element pattern (square patch) to determine the results. In the measurement, the cluster scattering from nearby objects is calibrated out to improve the measuring accuracy, specifically the sidelobe levels. In a word, the measured radiation characteristics in Figs 4.17 and 4.18 clearly demonstrate the dual-mode operation capability of the integrated beam-switching/retrodirective array.

4.3.5 *Complementary Design*

The dual-mode phased array in Sec 4.3.4 is not the only possible solution to integrate two different array structures as a whole. To illustrate this, an alternative form using dual-mode left-handed synthesized lines in Chapter 3 is introduced in the following. The response of the feeding network is a complement to that of the original design. Again, without the need of any active switches such as the PIN diode or MEMS, it can switch the circuit topology between two operational modes: in the low band, it resembles a Van Atta array, while in the high band it is a Butler matrix with multiple selectable inputs.

Figure 4.19(a) illustrates the schematic diagram of the complementary feeding network for realizing a dual-mode retrodirective/beam-switching array. Again, its architecture is based on

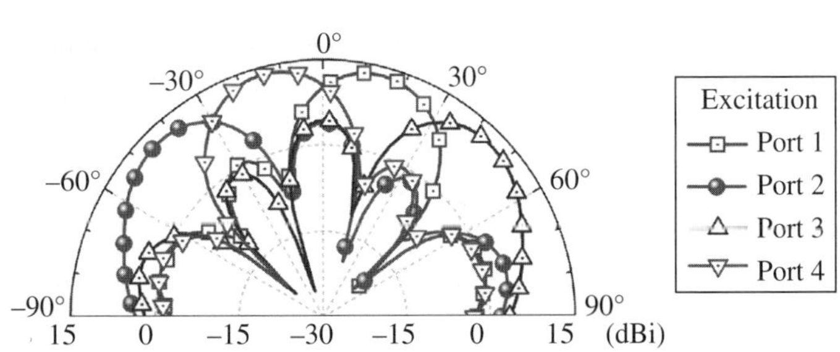

Figure 4.17 Measured radiation patterns of the heterogeneous integrated phased array in the beam-switching mode as ports 1–4 are excited alternatively. Source: Zou 2013 [24]. Reproduced with permission of IEEE

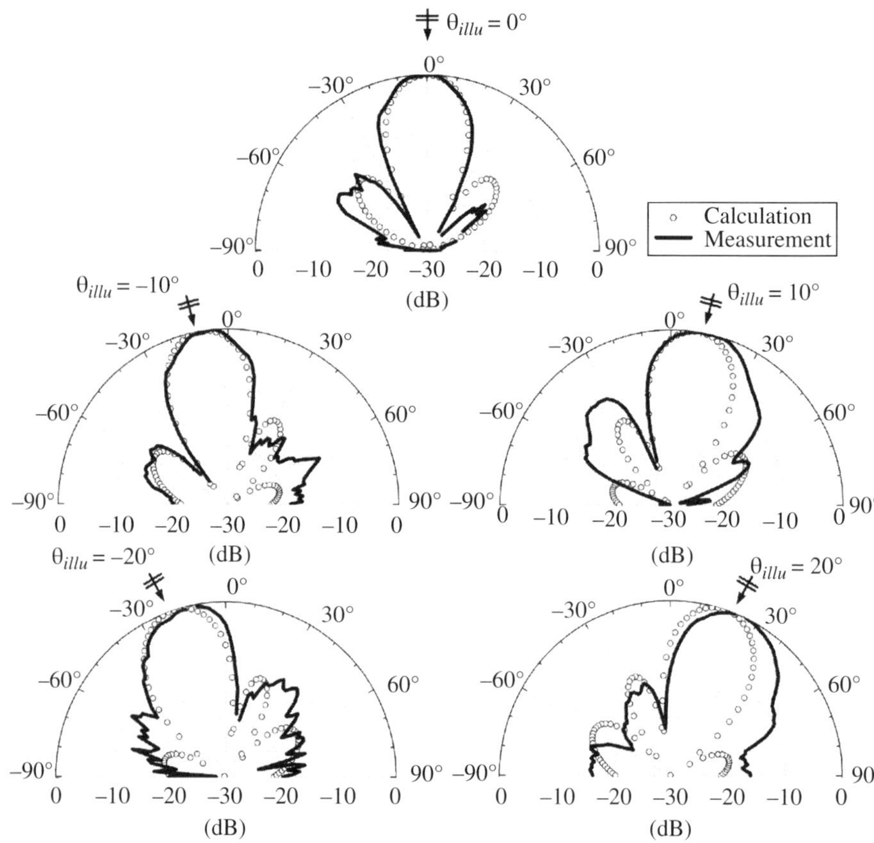

Figure 4.18 Calculated and measured bistatic radiation patterns of the heterogeneous integrated phased array in the RDA mode. Source: Zou 2013 [24]. Reproduced with permission of IEEE

the conventional four-element Butler matrix, and is composed of a dual-mode crossover and two dual-mode couplers. The dual-mode coupler is equivalent to a pair of isolated transmission lines oriented *in the shunt paths* in the low band, but behaves identical to a traditional 3-dB quadrature coupler in the high band. The dual-mode crossover, which is a cascade connection of two dual-mode couplers with a pair of phase adjustment lines, has similar characteristics as well. Since the dual-mode coupler is able to prevent any signal from passing through its serial paths in the RDA mode such that the design is equivalent to a Van Atta array, no dual-mode delay lines are required at the input terminals. The couplers and crossover in the dot frame are single-band designs centered at f_H for the same reason. The circuit layout of the feeding network is shown in Fig. 4.19(b). The signal flows in the low band (as a Van Atta array) are plotted for easy understanding. Again, suitable phase compensation is required for proper operations in both bands.

At 2.4 GHz (low band), the retransmitted radiation patterns of the integrated phased array are measured bistatically and presented in 4.20(a). The incident angle of the illuminating signal varies from −20° to +20° with a 10° step. The incoming signals are successfully retransmit to the direction of arrival using this alternative design worked in the RDA mode.

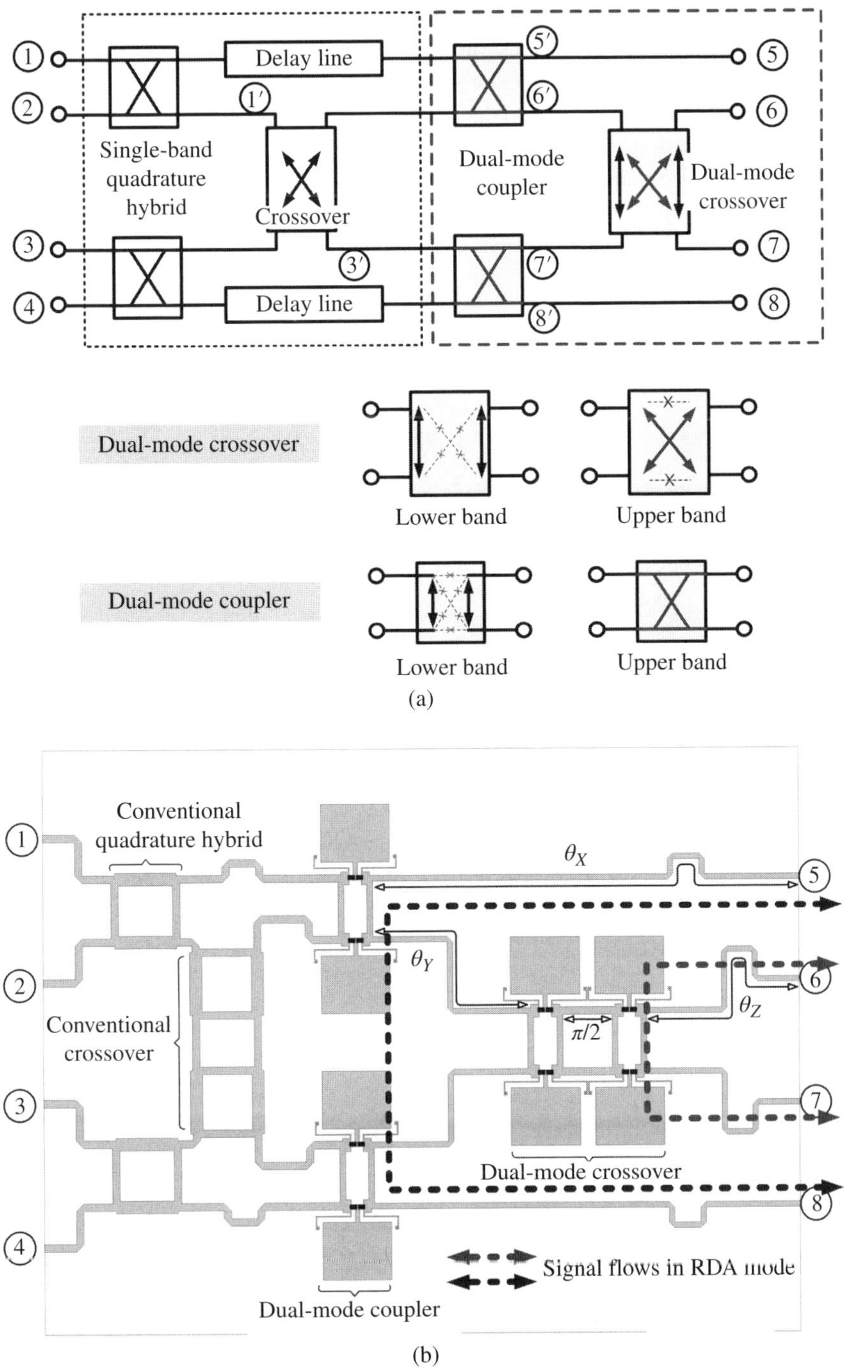

Figure 4.19 (a) Schematic diagram and (b) circuit layout of the dual-mode feeding network for a heterogeneous integrated retrodirective/beam-switching array

Figure 4.20(b) further depicts the calculated and measured radiation patterns when the dual-mode integrated array is operated in the beam-switching mode (5.15 GHz). Since the feeding network is identical to a typical 4×4 Butler matrix, the main beam is steered to −46°, −14°, 14°, and 46° like a conventional design.

A number of variations, such as the integration of a reflection-type RDA and a beam-switching array, can be realized in a similar way. Interested readers are encouraged to brainstorm and come up with completely new ideas!

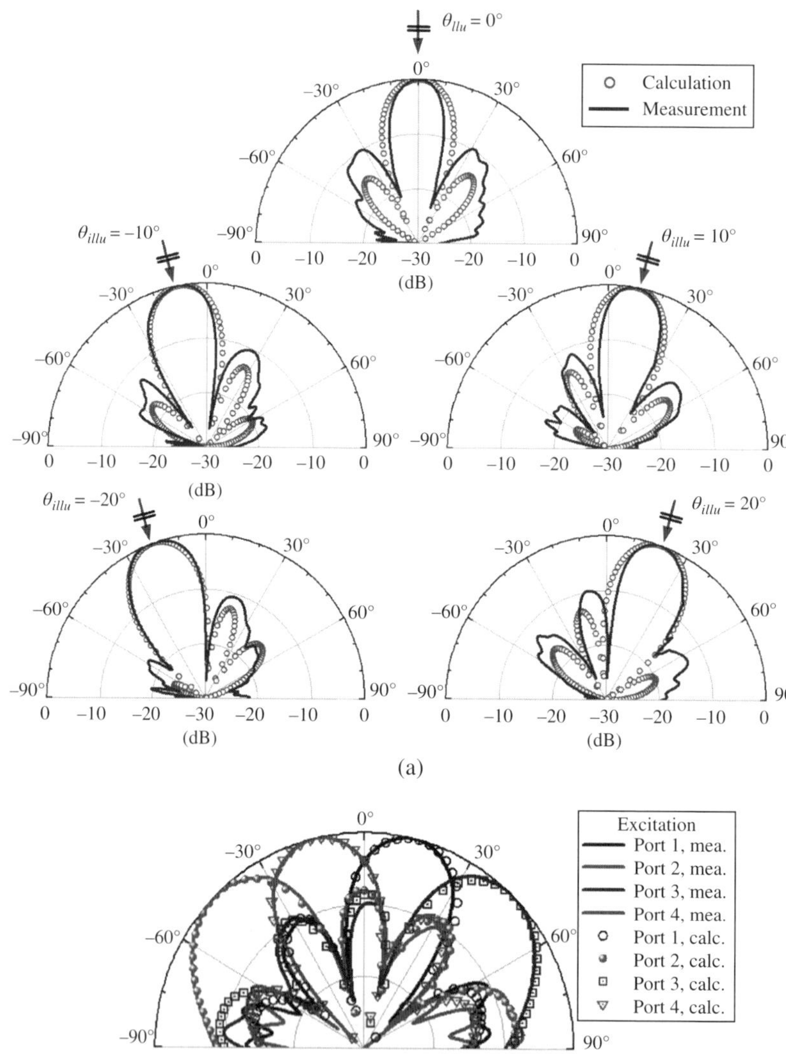

Figure 4.20 (a) Bistatic patterns in the RDA mode and (b) radiation patterns in the beam-switching mode of the integrated retrodirective/beam-switching array

4.4 Tri-Mode Heterogeneous Integrated Phased Array

4.4.1 Design Goal

In this section, the concept of dual-mode operation is further extended to design a tri-mode heterogeneous integrated phased array using tri-operational mode synthesized lines presented in Chapter 3. The integrated phased array can automatically reconfigure its feeding topology into a beam-switching array, a Van Atta array, and a phase-conjugating array, respectively, in the low, mid, and high bands [26]. The application scenario is illustrated in Fig. 4.21.

Referring to Fig. 4.21(a), the feeding network is identical to a beam-switching array in the low band, capable of providing full-duplex communication channels between the array and ambient devices. Its beam-switching capability, meanwhile, enables geometrical positioning. In the mid band, benefiting from the tri-mode building blocks, the integrated array automatically changes its topology to a Van Atta array. Together with extra bi-directional amplifiers, the integrated array may enable wireless power transmission and/or wireless charging to surroundings (Fig. 4.21(b)). In the high band, meanwhile, the integrated array again shows its self-adaptive ability and becomes a phase-conjugating array (PCA) as shown in Fig. 4.21(c). By injecting a slow-varying modulated signal into the local oscillator, the integrated array can function similar to an active transponder which can transfer its own information such as the temperature and humidity back to the interrogator.

The array, able to actively or passively exchange information, is kind of *smart*, and may find applications in smart living space, wireless sensor network, and even large-scale public demonstration as an agile communicator.

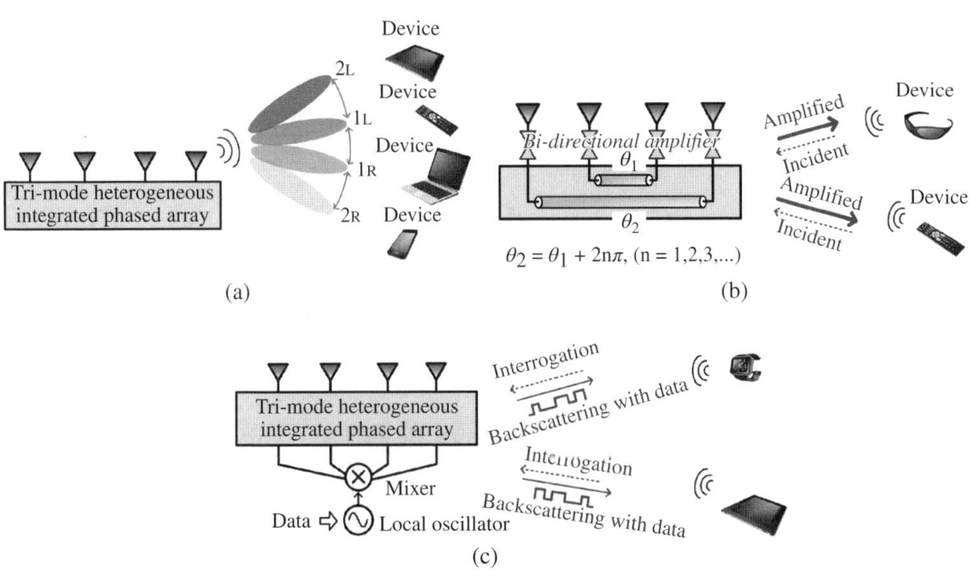

Figure 4.21 Application scenario of the heterogeneous integrated phased array in the (a) low band, (b) mid band, and (c) high band. Source: Lai 2014 [26]. Reproduced with permission of IEEE

4.4.2 System Architecture

Figure 4.22(a) shows the schematic diagram of the feeding network for realizing a tri-mode heterogeneous integrated beam-switching/Van Atta/PCA array. Here, we again focus on a four-element array. This particular design, offering dramatically different functionalities at three frequencies, is mainly composed of three passive components, the three-channel selector,

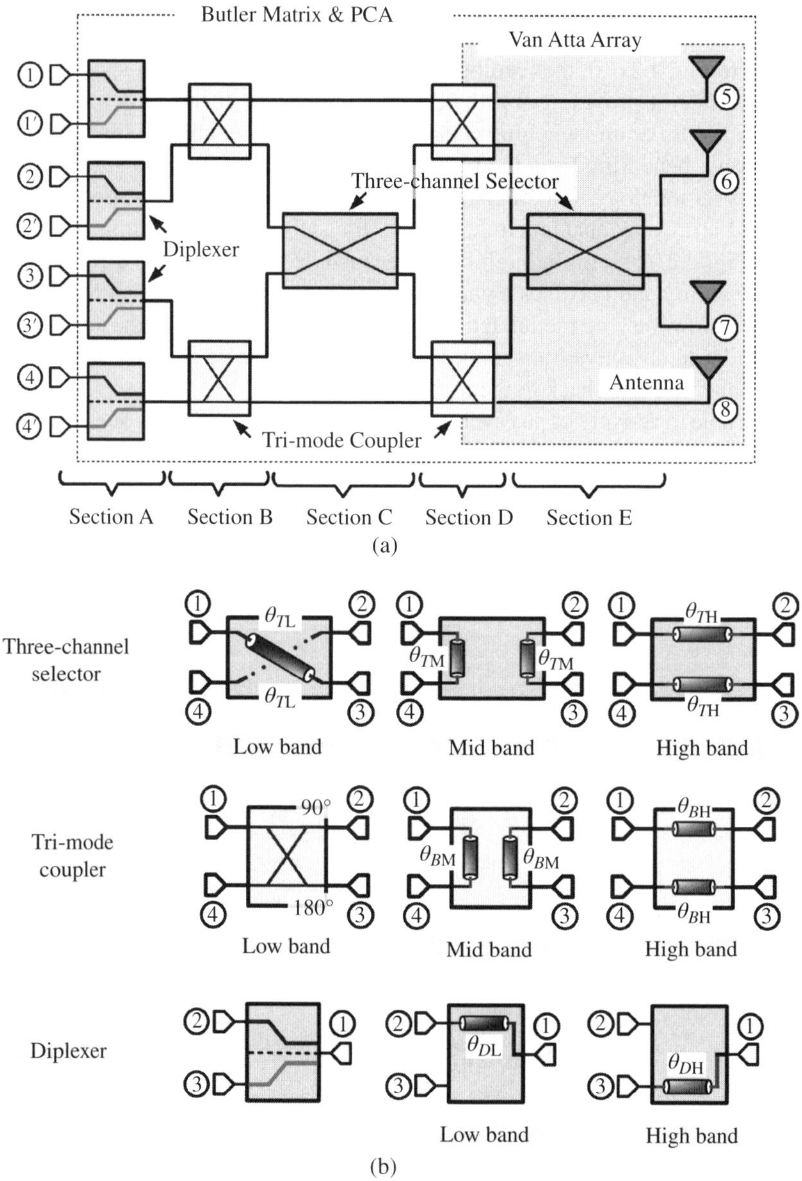

Figure 4.22 (a) Schematic diagram and (b) operational states of the core components of the tri-mode heterogeneous integrated phased array. Source: Lai 2014 [26]. Reproduced with permission of IEEE

tri-mode coupler, and diplexer. The operational states of the key components are summarized in Fig. 4.22(b)

The most important building block is the *three-channel selector*, which is, at the first glance, very similar to the triplexer in Sec. 3.5.2. Nevertheless, due to its two-fold symmetry, each of the four ports can serve as the common input [27]. When one of the ports is excited, the remaining ports of the channel selector either function as a corresponding transmission channel with good impedance matching, or provide very high isolation to the input. Referring to Fig. 4.22(b), the three-channel selector is equivalent to a conventional crossover in the low band, but forms pairs of isolated 50-Ω delay lines in the mid and high bands. The signals are delivered, spatially, to orthogonal directions in these two bands. In the channel selector, the electrical lengths of the transmission paths are respectively denoted as θ_{TL}, θ_{TM}, and θ_{TH} in the operating bands.

The tri-mode coupler, having the same design basis as the channel selector, has similar responses except that in the low band, it is a typical quadrature hybrid coupler. Also in Fig. 4.22(b), the electrical lengths of the direct-thru paths in the mid and high bands are marked as θ_{BM} and θ_{BH}, respectively.

Finally, the diplexer is used to separate the low- and high-band signals at the common inputs (port 1–4 and 1'–4'). Its electrical length is θ_{DL} in the low band, and θ_{DH} in the high band. It is designed by dual-mode left-handed/right-handed synthesized lines in Chapter 3.

4.4.3 Operation and System Implementation

The operation of the heterogeneous integrated phased array can be further examined using the signal flows in Fig. 4.23(a)–(c). The feeding network is identical to a typical 4×4

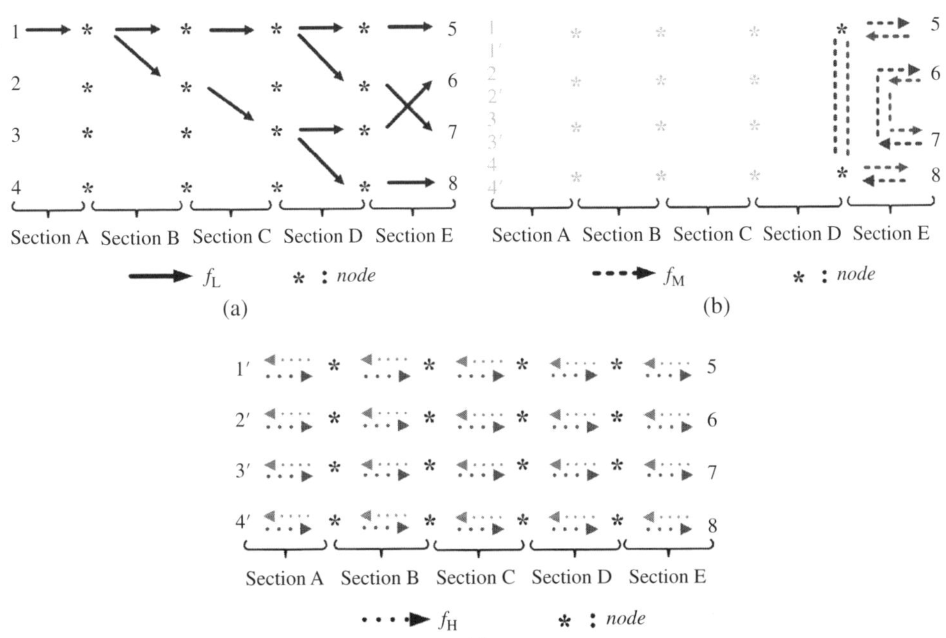

Figure 4.23 Signal flows of the feeding network for the tri-mode integrated array in the (a) low, (b) mid, and (c) high bands. Source: Lai 2014 [26]. Reproduced with permission of IEEE

Butler matrix consisting of couplers and crossovers in the low band (f_L). The inputs are ports 1–4 while the outputs are ports 5–8. The signal flow, with the port 1 being excited, is given in Fig. 4.23(a). The desired progressive phase difference between adjacent radiating elements, for switching the beam, can be easily achieved using the couplers and crossovers.

In the mid band (f_M), the feeding network automatically reconfigures its circuit topology and functions identical to a four-element Van Atta array. With the channel selector and tri-mode coupler both operated in the direct-thru mode, only the shadow area in Fig. 4.22(a) is activated; it forms two independent transmission channels (path 5–8 and path 6–7) in the vertical direction. The antennas are therefore connected in pairs. To fulfill the phase difference reversal condition for retrodirection, the relative phase between the two paths should be an integer multiple of 360°.

The feeding network in the high band (f_H) provides four independent transmission channels (paths 1'–5, 2'–6, 3'–7, and 4'–8) with high isolation in-between. The phase difference reversal condition is achieved by connecting the ports 1'–4' to down-conversion mixers. The signals are mixed by a common local oscillator to fulfill the operational requirement of a phase conjugating array (PCA) with heterodyne conversion.

Similar to the designs in Sec. 4.3, in order to simultaneously fulfill the required phase conditions of the beam-switching, Van Atta, and phase-conjugating arrays, a phase compensation scheme using multiple connecting lines is necessary to account for the frequency-dependent phase delays of the tri-mode components in the three bands. Figure 4.24(a)–(c) provides a comprehensive illustration of the phase relationships in the three states. The connecting segments for compensation purpose, $\theta_1 \sim \theta_6$, are shown at the same time. The design formulae, after manipulation, are summarized as:

$$\left(\theta_1 + \theta_3 + \theta_{TL}\right) - \theta_2 = 2n_1\pi, \quad n_1 = 0, \pm 1, \cdots \tag{4.4}$$

$$\theta_3 = {n_2\pi}/{2k_M} + n_1\pi - {\theta_{BM}}/{2k_M} - {\theta_{TL}}/{2} \tag{4.5}$$

$$\theta_4 = \left(2\pi - \theta_{TL}\right)/2 \tag{4.6}$$

$$\theta_6 = \frac{1}{k_H}\left(2n_3\pi - \left(k_H\theta_1 - k_H\theta_2 + k_H\theta_3 + 2k_H\theta_4 - k_H\theta_5 + 2\theta_{TH}\right)\right) \quad n_3 = 0, \pm 1, \cdots. \tag{4.7}$$

$$k_M = f_M / f_L; k_H = f_H / f_L. \tag{4.8}$$

In (4.4), since there is only one equation but two variables (θ_1, θ_2), one has a degree of freedom that can be determined in advance. Meanwhile, the segment θ_5, like the delay lines in a conventional Butler matrix, is simply 45° at the lowest operation frequency. The integers n_1, n_2, n_3

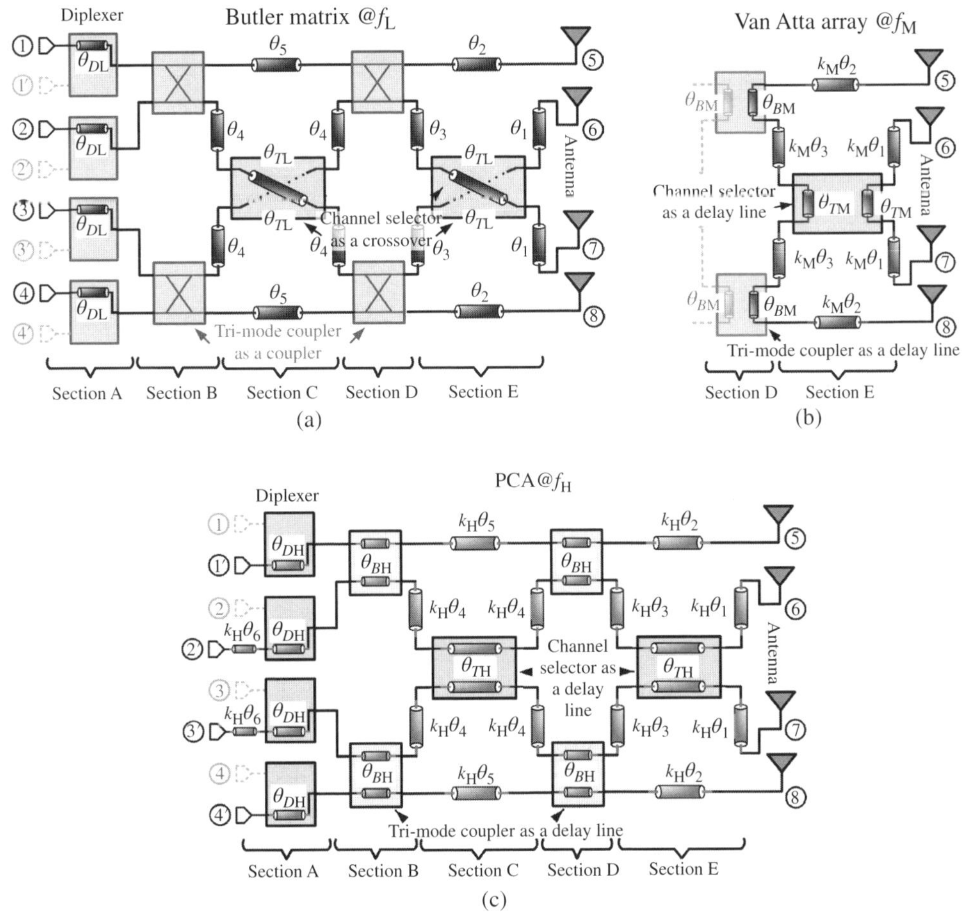

Figure 4.24 Phase relationship of the tri-mode feeding network in the (a) low, (b) mid, and (c) high bands. Source: Lai 2014 [26]. Reproduced with permission of IEEE

are selected in accordance with practical layout considerations. The interested readers may find a detailed derivation in [27].

4.4.4 Circuit Responses and Radiation Patterns

By applying the multi-mode synthesized transmission lines in Chapter 3, the tri-mode feeding network is realized and the circuit layout is shown in Fig. 4.25(a). Blocks A, B, and C in Sec. 3.5.2 are adopted to complete the design. Fig. 4.25(b) further illustrates the block diagram of the whole system including the gain amplifiers, circulators, radiators, and so on. The operating bands are selected as 0.7, 1.85, and 3.55 GHz. The circuit performances of the feeding network as well as the radiation characteristics of the integrated array are introduced and discussed successively in the following.

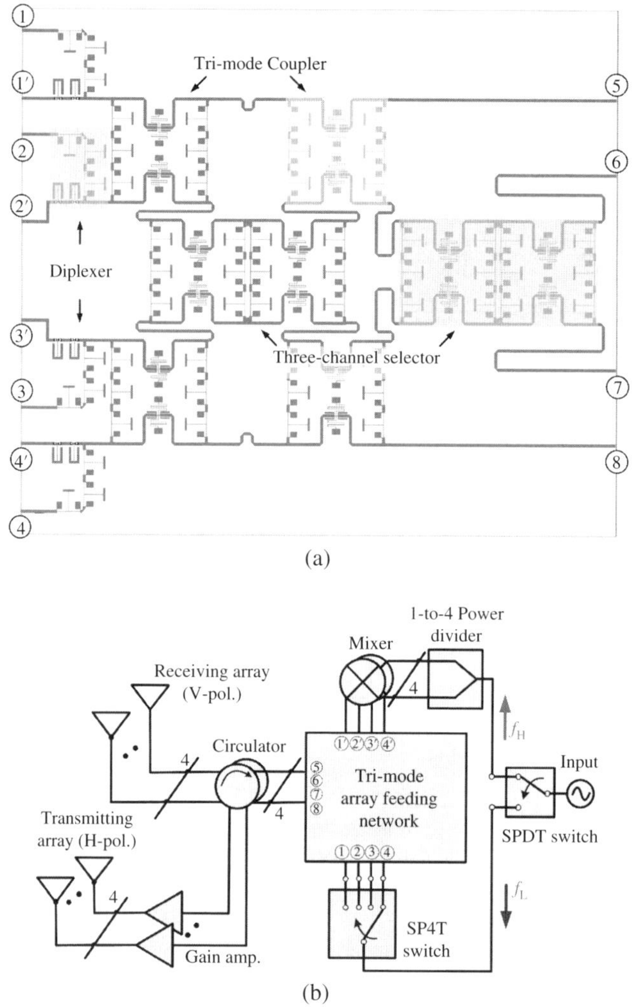

Figure 4.25 Tri-mode heterogeneous integrated phased array: (a) circuit layout of the feeding network and (b) block diagram of the whole system. Source: Lai 2014 [26]. Reproduced with permission of IEEE

4.4.4.1 Beam-Switching Mode

In the low band, to make the heterogeneous integrated phased array functions as a beam-switching one, the SPDT directs the signal to the lower path of the network as illustrated in Fig. 4.25(b). The SP4T, meanwhile, selects the radiation beam via the input channels (ports 1 ~ 4). The receiving array as well as the mixing circuit is inactive in this mode. The transmission coefficients of the feeding network are −6.7 ~ −8.1 dB at 0.7 GHz. The isolations between channels are better than 30 dB. The maximum error of the relative phases between adjacent outputs is less than ±3°.

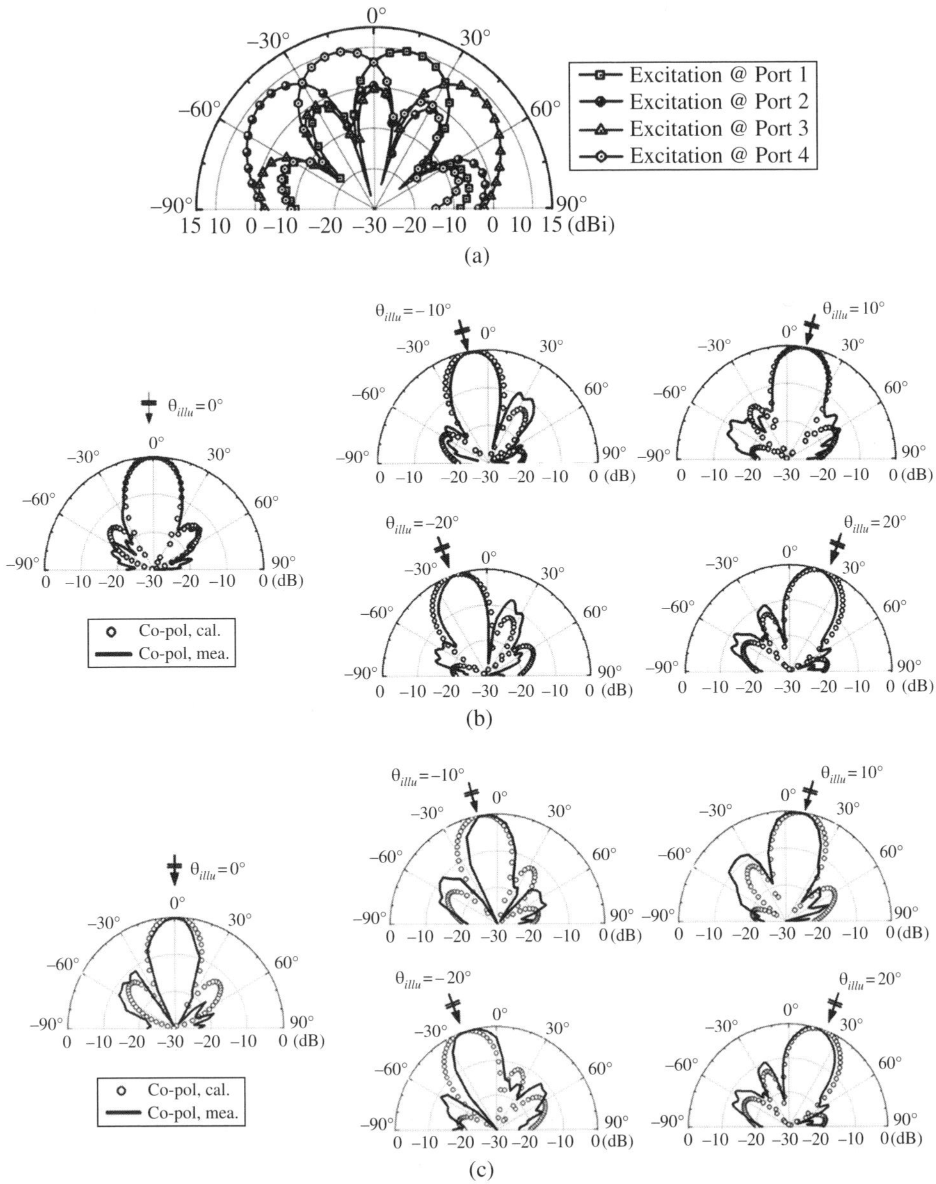

Figure 4.26 Measured bistatic radiation patterns of the tri-mode integrated array in the (a) beam-switching mode at 0.7 GHz, (b) Van Atta mode at 1.85 GHz, and (c) PCA mode at 3.55 GHz. Source: Lai 2014 [26]. Reproduced with permission of IEEE

Figure 4.26(a) depicts the measured radiation patterns of the tri-mode integrated phased array at 0.7 GHz. The experimental results are very close to the theoretical ones. Upon the selection of the input channels (ports 1 ~ 4), the beam is steered to 14°, −45°, 45°, and −14°. The measured peak gains are 9.9, 8.4, 8.7, and 9.9 dBi.

4.4.4.2 Van Atta Mode

The integrated phased array reconfigures its topology and becomes a Van Atta array in the mid band (f_M). Referring to Fig. 4.25(b), the incident signal captured by the receiving array is fed into the ports 5–8 of the feeding network via circulators. The network is equivalent to a pair of isolated transmission lines whose phase difference is an integer multiple of 2π. The reentrant signal into the circulators hence satisfies the phase difference reversal condition, and re-radiates toward to the direction of arrival using the transmitting array. The ports 1 ~ 4 and 1' ~ 4', in the meantime, are isolated from the active part of the circuit.

At this frequency (1.85 GHz), the measured transmission coefficients of the feeding network ($|S_{85}|$ and $|S_{76}|$) are −3 and −1.8 dB, and the phase difference between the paths is around 10°. The amplitude imbalance is a result of different numbers of components in the transmission paths, while the phase imbalance (other than 0°) can be accounted for by the accumulated errors of the nonideal circuit elements. The isolations are around 60 dB.

The retransmitted radiation patterns of the integrated array in the Van Atta mode are measured bistatically at 1.85 GHz. The calculated and measured bistatic (retransmitted) radiation patterns at 0°, ±10°, and ±20° are summarized in Fig. 4.26(b). The array clearly demonstrates its retrodirective beam steering capability via the Van Atta configuration in the mid band.

4.4.4.3 PCA Mode

Referring to Fig. 4.25(b), in the PCA mode the LO signal is injected and evenly distributed to the mixers using the SPDT and 1-to-4 power divider. The tri-mode feeding network, in the meantime, creates four independent transmission channels and guides the RF signals from the receiving antennas to pass through the network and arrive at the gate terminals of the mixers. The nonlinear mixers then generate IF signals that comply with the phase difference reversal condition. The signals are retransmitted back to the transmitting array via the circulators and gain amplifiers. It completes the retrodirection process.

At 3.55 GHz, the measured transmission coefficients $|S_{51'}|$ and $|S_{62'}|$ of the feeding network are −3.5 and −6.5 dB; the phase difference deviates from the ideal value by only 5°. The amplitude and phase imbalances are caused by similar reasons observed in the Van Atta mode, and can be compensated for by adjusting the individual gain stage in the transmission path. The isolations, in the meantime, are higher than 48 dB.

The retransmitted radiation patterns of the tri-mode array in the PCA mode are measured and summarized in Fig. 4.26(c). Two signal generators are used to provide the LO and RF signals at the inputs of the phased array and illuminating antenna, respectively. The LO and RF frequencies are 7.1 and 3.54 GHz. The incident angle of the illuminating wave varies from −20° to +20° with a 10° step. The radiation patterns are recorded manually from a spectrum analyzer every 10° in the azimuth plane. The retransmitted patterns again validate the successful implementation of the tri-mode heterogeneous integrated array.

4.5 Simplified Dual-Mode Integrated Array Using Two Elements

Some readers may think the dual-/tri-mode integrated phased arrays introduced in previous sections are too complicated to be realized. Indeed, in certain applications with a limited mounting space or payload, the feeding network of the four-element array could be too large

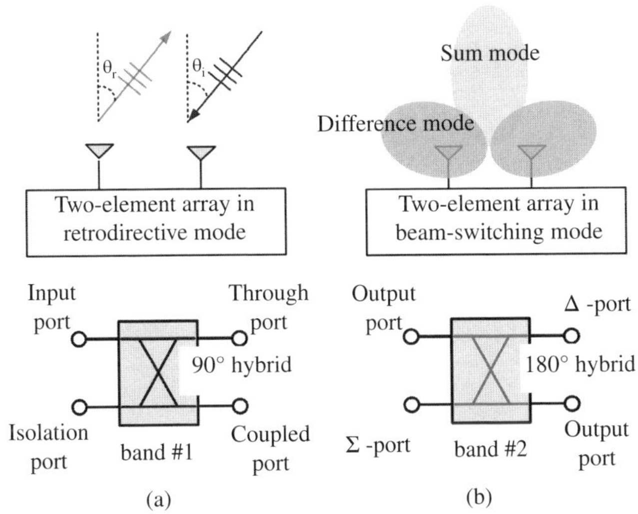

Figure 4.27 (a) Design goal and (b) key component of a two-element dual-mode integrated retrodirective/beam-switching array

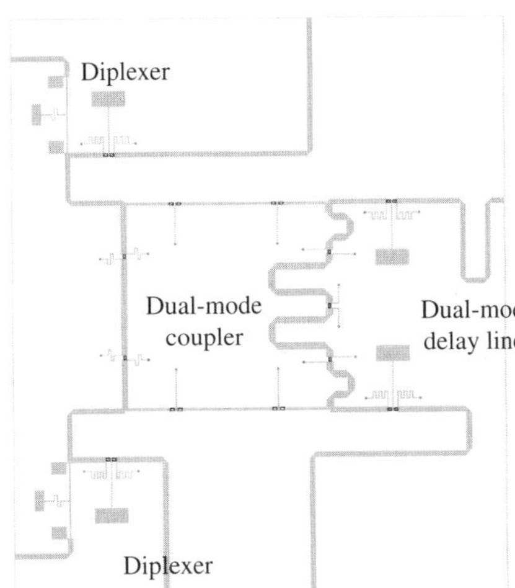

Figure 4.28 Circuit layout of the two-element dual-mode integrated retrodirective/beam-switching array

to be installed. To tackle the problem and to demonstrate the flexibility of synthesized transmission lines, in this section a two-element dual-mode integrated array is briefly discussed.

Figure 4.27(a) shows the design goal of the two-element array. In the band #1, it is again a retrodirective array while in the band #2 it becomes a beam-switching one with sum and difference patterns. As indicated in Fig. 4.27(b), the key component in the circuit realization is a dual-mode hybrid coupler. Differing from those previously discussed, this dual-mode design should be equivalent to a 90° and a 180° hybrid coupler respectively in the two bands. The

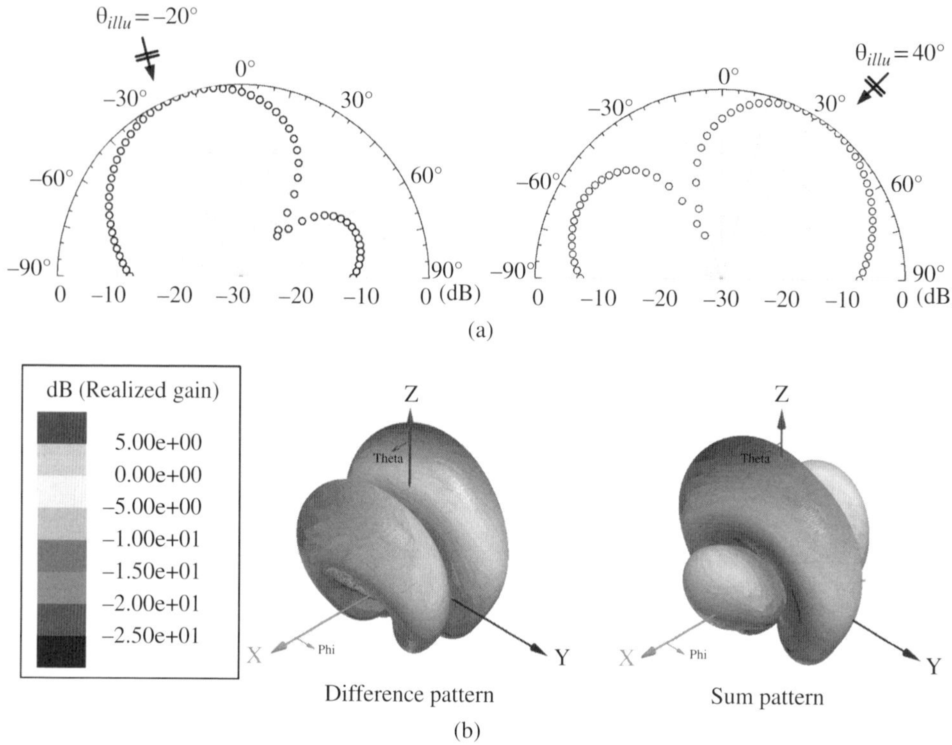

Figure 4.29 Radiation patterns of the two-element dual-mode integrated array in (a) band #1 (retrodirection) and (b) band #2 (beam-switching)

implementation is also relied on the dual-mode synthesized transmission lines. Nevertheless, here the characteristic impedance and phase delay of the dual-mode lines must be taken into accounted simultaneously at both operating frequencies. This brings extra design challenges.

Figure 4.28 provides a complete layout of the feeding network for the two-element retrodirective/beam-switching array. In addition to the core component, that is, the dual-mode hybrid coupler, it consists of diplexers and dual-mode delay lines, all realized using synthesized transmission lines. The radiation patterns in the band #1 (retrodirection) and band #2 (beam-switching) are shown in Fig. 4.29(a) and (b). Since the array is composed of only two elements, the scanning range, i.e. the beamwidth in which the RDA functions properly, is considerably extended to at least ±40°. In the meantime, the sum and difference patterns in the band #2 are completely orthogonal to each other, with the enveloped correlation coefficient (ECC) close to zero.

References

[1] J. Butler and R. Lowe, "Beam forming matrix simplifies design of electronically scanned antennas," *Electron. Design*, vol. 9, pp. 170–173, Apr. 1961.

[2] M. Bona, L. Manholm, J. P. Starski, and B. Svensson, "Low-loss compact Butler matrix for a microstrip antenna," *IEEE Trans. Microw. Theory Techn.*, vol. 50, no. 9, pp. 2069–2075, Sep. 2002.

[3] H. Hayashi, D. A. Hitko, and C. G. Sodini, "Four-element planar Butler matrix using half-wavelength open stubs," *IEEE Microw. Wireless Compon. Lett.*, vol. 12, no. 3, pp. 73–75, Mar. 2002.

[4] M. Nedil, T. A. Denidni, and L. Talbi, "Novel Butler matrix using CPW multilayer technology," *IEEE Trans. Microw. Theory Techn.*, vol. 54, no. 1, pp. 499–507, Jan. 2006.

[5] Y. S. Jeong and T. W. Kim, "Design and analysis of swapped port coupler and its application in a miniaturized Butler matrix," *IEEE Trans. Microw. Theory Techn.*, vol. 58, no. 4, pp. 764–770, Apr. 2010.

[6] K. Wincza, S. Gruszczynski, and K. Sachse, "Broadband planar fully integrated 8×8 Butler matrix using coupled-line directional couplers," *IEEE Trans. Microw. Theory Techn.*, vol. 59, no. 10, pp. 2441–2446, Oct. 2011.

[7] C.-C. Chang, R.-H. Lee, and T.-Y. Shih, "Design of a beam switching/steering Butler matrix for phased array system," *IEEE Trans. Antennas Propagat.*, vol. 58, no. 2, pp. 367–374, Feb. 2010.

[8] T.-Y. Chin, S.-F. Chang, J.-C. Wu, and C.-C. Chang, "A 25-GHz compact low-power phased-array receiver with continuous beam steering in CMOS technology," *IEEE J. Solid-State Circuits*, vol. 45, no. 11, pp. 2273–2282, Nov. 2010.

[9] S. Mosca, F. Bilotti, A. Toscano, and L. Vegni, "A novel design method for Blass matrix beam-forming networks," *IEEE Trans. Antennas Propagat.*, vol. 50, no. 2, pp. 225–232, Feb. 2002.

[10] N. J. G. Fonseca, "Printed *S*-band 4×4 Nolen matrix for multiple beam antenna applications," *IEEE Trans. Antennas Propagat.*, vol. 57, no. 6, pp. 1673–1678, Jun. 2009.

[11] L. C. Van Atta, "Electromagnetic reflector," U.S. patent 2 908 002, Oct. 6, 1963.

[12] R. Y. Miyamoto and T. Itoh, "Retrodirective arrays for wireless communications," *IEEE Microwave Magazine*, vol. 3, no. 1, pp. 71–79, Mar. 2002.

[13] W.-J. Tseng, S.-J. Chung, and K. Chang, "A planar Van Atta array reflector with retrodirectivity in both E-plane and H-planes," *IEEE Trans. Antennas Propagat.*, vol. 48, no. 2, pp. 173–175, Feb. 2000.

[14] S.-J. Chung, S.-M. Chen, and Y.-C. Lee, "A novel bi-directional amplifier with applications in active Van Atta retrodirective arrays," *IEEE Trans. Microw. Theory Techn.*, vol. 51, no. 2, pp. 542–547, Feb. 2003.

[15] C. Y. Pon, "Retrodirective array using the heterodyne technique," *IEEE Trans. Antennas Propagat.*, vol. 12, pp. 176–180, Mar. 1964.

[16] R. Y. Miyamoto, Y. Qian, and T. Itoh, "An active integrated retrodirective transponder for remote information retrieval-on-demand," *IEEE Trans. Microw. Theory Techn.*, vol. 49, no. 9, pp. 1658–1662, Sep. 2001.

[17] L. Chiu, Q. Xue, and C. H. Chan, "Phase-conjugated arrays using low conversion-loss resistive phase-conjugating mixers and stub-loaded patch antennas," *IEEE Trans. Microw. Theory Techn.*, vol. 56, no. 8, pp.1764–1773, Aug. 2008.

[18] T. Brabetz, V. F. Fusco, and S. Karode, "Balanced subharmonic mixers for retrodirective-array applications," *IEEE Trans. Microw. Theory Techn.*, vol. 49, no. 3, pp. 465–469, Mar. 2001.

[19] S.-C. Yen and T.-H. Chu, "A retro-directive antenna array with phase conjugation circuit using subharmonically injection-locked self-oscillating mixers," *IEEE Trans. Antennas Propagat.*, vol. 52, no. 1, pp. 154–164, Jan. 2004.

[20] S.-N. Hsieh and T.-H. Chu, "Linear retro-directive antenna array using 90° hybrids," *IEEE Trans. Antennas Propagat.*, vol. 56, no. 6, pp. 1573–1580, Jun. 2008.

[21] J.-W. Tsai, C.-H. Wu, and T.-G. Ma, "Novel dual-mode retrodirective array using synthesized microstrip lines," *IEEE Trans. Microw. Theory Techn.*, vol. 59, no. 12, pp. 3375–3388, Dec. 2011.

[22] J.-W. Tsai and T.-G. Ma, "Compact dual-mode four-port network with quadrature-coupling and direct-thru transmission in each of the individual bands", in *2011 IEEE International Microw. Symp. Dig.* Baltimore MD, Jun. 2011.

[23] P. Bura and R. Dikshit, "FET mixers for communication satellite transponders", in *IEEE MTT-SInt. Microw. Symp. Dig.*, 1976, pp. 90–92.

[24] J.-Y. Zou, C.-H. Wu, and T.-G. Ma, "Heterogeneous integrated beam-switching/retrodirective array using synthesized transmission lines," *IEEE Trans. Microw. Theory Techn.*, vol. 61, no. 8, pp. 3128–3139, Aug. 2013.

[25] H. Kim, B. Lee, and M.-J. Park, "Dual-band branch-line coupler with port extensions," *IEEE Trans. Microw. Theory Techn.*, vol. 58, no. 3, pp. 651–655, Mar. 2010.

[26] C.-H. Lai, C.-Y. Shiau, and T.-G. Ma, "Tri-mode heterogeneous integrated beam-switching/Van Atta/phase-conjugating array using synthesized transmission lines," *IEEE Trans. Microw. Theory Techn.*, vol. 62, no. 9, pp. 2180–2192, Sep. 2014.

[27] C.-H. Lai, C.-Y. Shiau, and T.-G. Ma, "Microwave three-channel selector using tri-mode synthesized transmission lines," *IEEE Trans. Microw. Theory Techn.*, vol. 61, no. 10, pp. 3529–3540, Oct. 2013.

5

On-Chip Realization of Synthesized Transmission Lines Using IPD Processes

Y. C. Tseng and T. G. Ma

5.1 Introduction

Mobile communications have experienced a blooming growth since the year of the new millennium. Portable devices such as smartphones, tablet PCs, and wearable devices are ubiquitously used by all ages of people around the world. Modern trends in mobile devices include compact size, multifunctional operation, low power consumption, robust to ambient environment, and so on. Due to limited circuit areas in mobile devices, there is an urgent demand for highly integrated digital/analog/RF modules and systems. Therefore, system-on-chip (SoC) and system-in-package (SiP) technologies have become the core business in communication industries over the past decade. While the SoC still faces a number of challenges today, the SiP technology is much more mature and has found practical applications recently. The passive microwave components, in the meantime, have become a major bottleneck in current system-in-package integration. The passive components always occupy a large chip area whose size is a function of frequency and difficult to scale down.

In previous chapters, a variety of synthesized transmission lines have been developed using the printed circuit board (PCB) technology with remarkable size reduction capability. Nevertheless, when the operating frequency goes high, one may experience difficulties in realizing synthesized transmission lines using PCB processes due to the limitation of line resolution. To alleviate the problem, the integrated passive device (IPD) process, having received considerable attention in the SiP integration, is applied to fulfill on-chip synthesized transmission lines and hence circuit miniaturization at higher frequencies. Benefitting from the improved

Synthesized Transmission Lines: Design, Circuit Implementation, and Phased Array Applications, First Edition.
Tzyh-Ghuang Ma, Chao-Wei Wang, Chi-Hui Lai and Ying-Cheng Tseng.

line resolution and low-loss properties, the IPD technology has been widely used in chip realization of passive components including couplers [1–6], balun [7], Butler matrix [8], bandpass filters [9–13], and multiplexers [14, 15]. Periodic synthesized structures such as the complementary-conducting strips [16], and non-periodic quasi-lumped networks with π/T-equivalences [1–3, 5, 6, 8] are the common building blocks for on-chip implementation of synthesized transmission lines.

In this chapter, a variety of synthesized transmission lines using the IPD technology are introduced to realize on-chip microwave/millimeter-wave passive components. On-chip synthesized lines, generally consisting of quasi-lumped meander/spiral inductors, metal-insulator-metal (MIM) capacitors, and open stubs, feature an extremely compact size with impressive slow wave factor. Various microwave components with acceptable power dissipation and good electrical responses are thereby realized. Tight couplers and bandpass/bandstop filters are among the topics to be covered. The concept of multi-mode operation in Chapter 3 is also fulfilled by quasi-lumped components on the IPD substrate. Their applications to the multiplexer, antenna decoupling network, and liquid detector are discussed.

5.2 Integrated Passive Device (IPD) Process

The IPD process, capable of providing higher-Q passive devices than their CMOS counterpart, is emerged as a competitive technology in the integration of on-chip microwave passive and active components. The typical profile of an IPD process is shown in Fig. 5.1. The carrier substrate could be glass or high-resistivity (high-R) silicon. It generally consists of three metal layers (M1, M2, and M3) with two dielectric layers stacked in between. The thicknesses of M1 and M2 layers are around 3 ~ 5 μm, while the M3 layer is an ultra-thick metal (UTM) whose thickness may exceed 10 μm. The thick metal layer is used to realize high-Q inductors in spiral or meander form, as depicted by the cross-sectional view in the figure. While the dielectric substrate between M2 and M3 layers is no more than a common low-loss material such as the Benzocyclobutene (BCB), the M1 and M2 metal layers are separated by a thin-film high-ε_r dielectric layer for realizing high-density MIM capacitors. Electrodes between the M3 and M1/M2 layers can be used to fulfill low-density MIM capacitors as well. The through via (M1-to-M3) is used to connect the top metal layer M3 to the bottom layer M1. Thin-film resistors are also possible with extra doping above the carrier substrate.

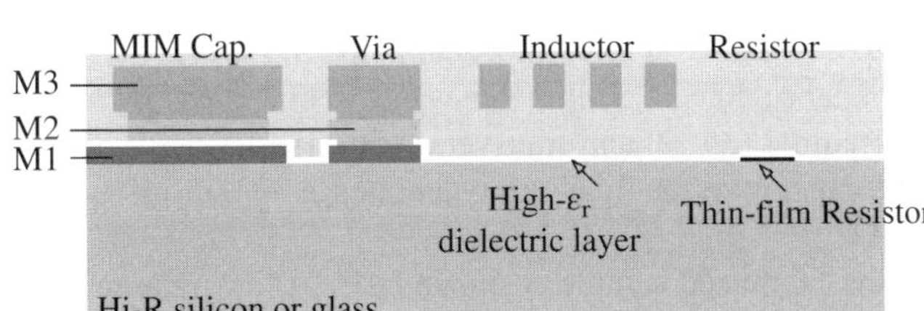

Figure 5.1 Cross-sectional profile of the IPD process

5.3 Tight Couplers Using Synthesized CPWs

Given the process profile, in this section, a number of on-chip tight couplers on the IPD substrate are introduced. Since the separation between the M3 (trace) and M1 (ground) layers is too narrow to realize common 50-Ω microstrip lines, most designs in the open literature are realized by means of CPWs and their associated quasi-lumped components. A V-band branch-line coupler, a wideband rat-race coupler, a dual-band rat-race coupler, and a directional coupler are discussed in succession.

5.3.1 Quadrature Hybrid

In the 50-Ω system, a conventional quadrature hybrid, or branch-line coupler, comprises two pairs of 90° lines whose characteristic impedances are respectively 50 and 35 Ω for equal power splitting. By utilizing the miniaturization technique in Sec. 1.4, a compact design with all signal traces equal to 70 Ω can be realized using non-periodic synthesized CPWs.

The schematic diagram of the synthesized CPW, or the so-called capacitively-loaded lower ground coplanar waveguide (LG-CPW), is given in Fig. 5.2(a) [1]. The original quarter-wavelength line is replaced by a high-impedance line loaded with two shunt capacitances. By equating the *ABCD* matrices of the uniform transmission line and the synthesized CPW, design equations are summarized as

$$Z_{0L} = Z_0 \frac{\sin\theta}{\sin\theta_L} \tag{5.1}$$

$$C = \frac{1}{\omega Z_0}\left(\frac{\cos\theta_L - \cos\theta}{\sin\theta}\right). \tag{5.2}$$

In (5.1), θ_L is a design variable which can be determined in advance to achieve an overall compact footprint. The shorter the length of the CPW section is, the higher its line impedance becomes. The required shunt capacitances rise accordingly to keep the characteristic impedance of the synthesized CPW at the desired value. As shown in Fig. 5.2(b), the schematic diagram is converted into the physical layout in CPW form. The capacitive loading is realized by shunt open-circuited stubs for the sake of precise control of the capacitance values. The ground plane of the CPW is allocated at the M1 layer (i.e., the lower ground) to achieve a high-impedance line at millimeter-wave frequencies. By setting the electrical lengths of the longitudinal (35-Ω) and transverse (50-Ω) CPW sections respectively equal to 30° and 45°, the capacitive loadings are calculated as 0.065 pF and 0.038 pF. All line impedances are 70 Ω.

A chip photo of the fabricated sample is shown in Fig. 5.3(a); the measured *S*-parameters are depicted in Fig. 5.3(b). The hybrid coupler is operated at the V-band and centered at 60 GHz. The chip size is $607 \times 722\ \mu m^2$. From 57 to 64 GHz, the phase error is less than 0.5° while the amplitude imbalance is lower than 1.2 dB. The transmission loss is 1.9 dB at the center frequency.

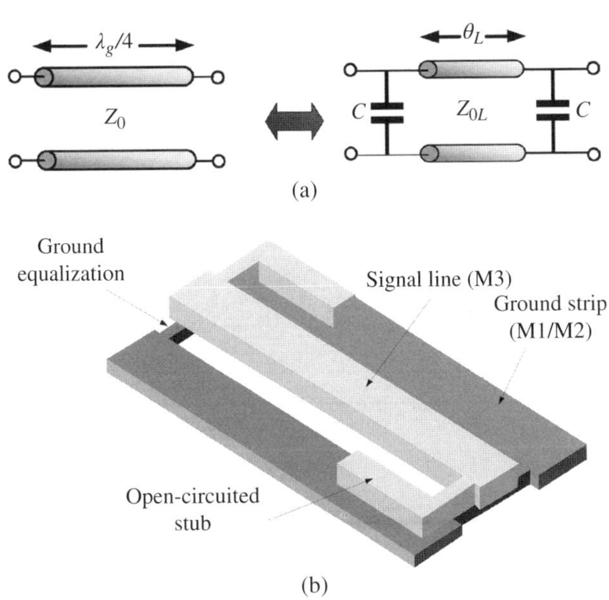

Figure 5.2 (a) Schematic diagram and (b) circuit layout of the capacitively-loaded lower ground coplanar waveguide (LG-CPW). Source: Haroun 2012 [1]. Reproduced with permission of IEEE

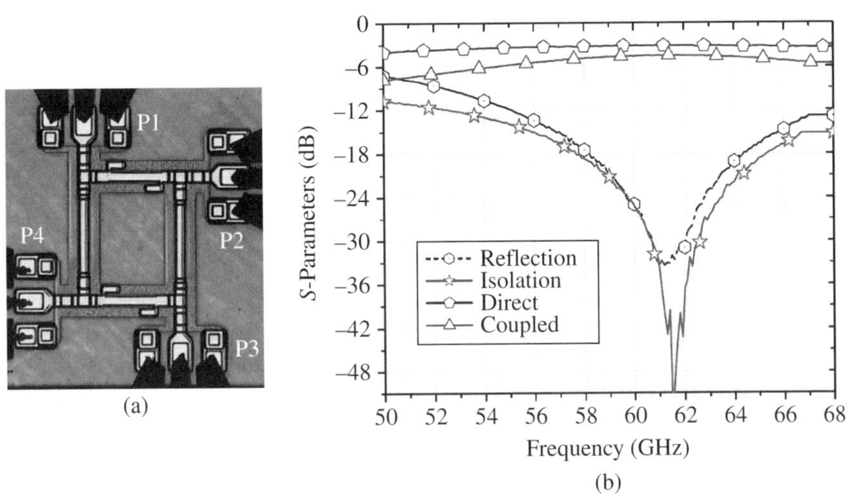

Figure 5.3 (a) Chip photo and (b) *S*-parameters of the V-band on-chip quadrature hybrid. Source: Haroun 2012 [1]. Reproduced with permission of IEEE

5.3.2 Wideband Rat-Race Coupler

In a conventional rat-race coupler, due to the existence of a three-quarter wavelength line, the bandwidth is generally limited to 15–20% as a narrowband design. To further widen the bandwidth, as proposed in [17, 18], one can introduce a phase inverter with a 180° phase shift between the ports. A 90° transmission line incorporated with the 180° phase inverter gives rise to a wideband equivalence of a three quarter-wavelength line.

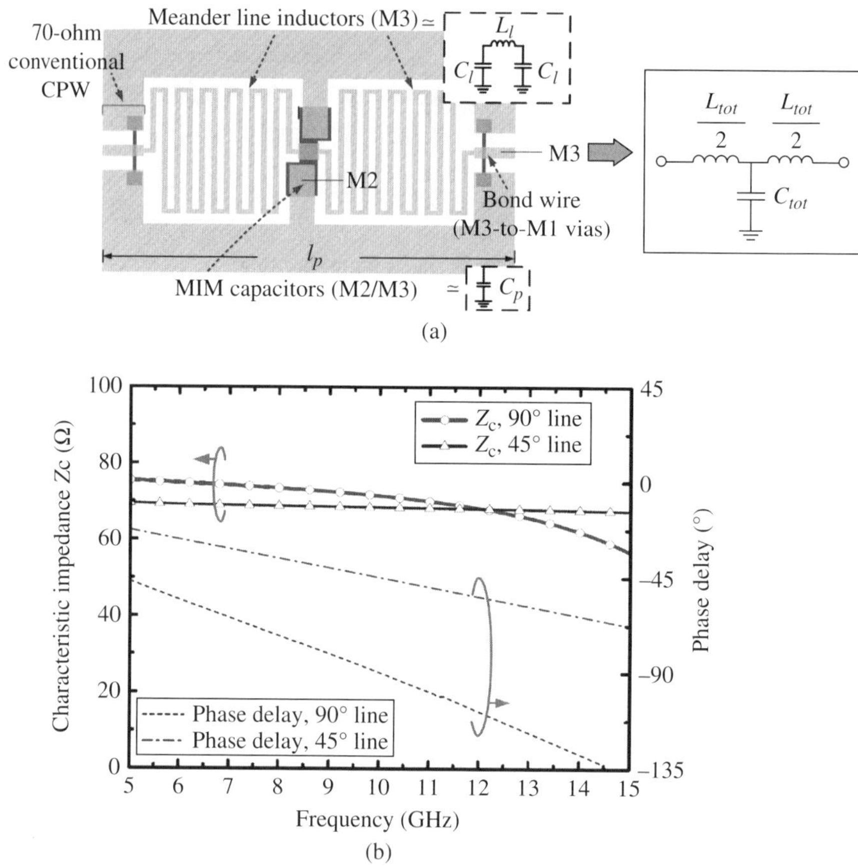

Figure 5.4 (a) Layout and (b) responses of the synthesized CPWs on the IPD substrate

With this basic principle in mind, an X-band on-chip wideband rat-race coupler is realized by five sections of slow-wave synthesized CPWs along with an additional phase inverter. Each quarter-wavelength line in a conventional ring coupler is replaced by a 70-Ω 90° synthesized CPW, while the 270° line is formed by two 70-Ω 45° synthesized CPWs in conjunction with a phase inverter. The layout of the synthesized CPW is shown in Fig. 5.4(a). It is a non-periodic synthesized line composed of two meander line inductors and MIM capacitors. The inductors, opposite to each other with respect to the center of the network, are designed on the thick metal layer M3 to minimize the conductor loss. The MIM capacitors, in the meantime, are realized by adjacent metal layers, M2 and M3. By utilizing the approximate technique of node exchange, the synthesized CPW is again modeled by a T-network consisting of two series inductances each with $L_{tot}/2 = L_l$ and a shunt capacitance $C_{tot} = 4C_l + 2C_p$. By equating the *ABCD* matrix of the synthesized CPW to that of a uniform transmission line, the characteristic impedance and electrical length are

$$Z_c = \sqrt{\frac{L_{tot}\left(1-\omega^2 L_{tot}C_{tot}/4\right)}{C_{tot}}} \tag{5.3}$$

$$\theta = \beta_g l_p = \cos^{-1}\left(1-\omega^2 L_{tot}C_{tot}/2\right). \tag{5.4}$$

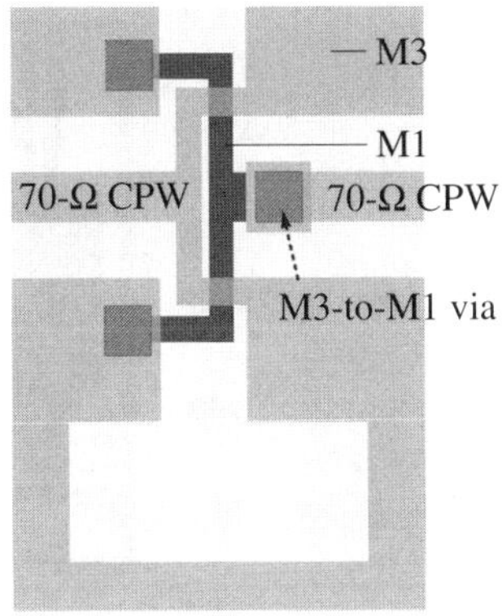

Figure 5.5 Layout of the phase inverter for wideband operation

The lumped parameters L_{tot} and C_{tot} can be solved using (5.3) and (5.4) by assigning Z_c and θ at the center angular frequency ω_0 of the operating band. In Fig. 5.4(a), the conventional 70-Ω CPW sections are used to adjust the electrical length of the line between the actual and target values. The bond wires, using M3-to-M1 vias, can suppress the higher-order modes. With the aid of the multilayer process, the MIM capacitors do not occupy any additional chip area, therefore making the synthesized CPW extraordinarily compact in size with a high slow wave factor of 8.5. The simulated characteristic impedances and electrical lengths of the 70-Ω 45° and 70-Ω 90° synthesized CPWs are plotted in Fig. 5.4(b).

Figure 5.5 shows the layout of the CPW-to-CPW phase inverter. To achieve a 180° phase reversal, the CPW-to-CPW transition should swap the signal line with the ground planes. For this purpose, the signal trace on the left side of the layout is connected directly to the ground planes of the CPW on the right side, and vice versa. The line interchange is accomplished using the M3-to-M1 vias and the connection lines in the M1 layer. To reduce the detrimental parasitic coupling, the lines in the M1 and M3 layers are offset by a small distance.

Ideally speaking, the inverter can achieve a perfect phase reversal when the CPWs on both sides of the inverter are electrically isolated to each other; however, in practice, the parasitic coupling between the lines degrades the isolation and hence the performance of the phase inverter. To tackle the problem, a compensated inductive rectangular slot is etched on one side of the ground planes as a quasi-lumped inductor [18]. This inductive slot and the parasitic capacitive coupling between the traces on the M1 and M3 layers form a parallel *LC* tank. The *LC* tank is nearly open-circuited at the center frequency of the coupler, which, in turn, improves the isolation and hence the performance of the phase inverter. The insertion loss of the phase inverter is less than 0.35 dB.

The circuit layout and chip photo of the miniaturized wideband rat-race coupler are shown in Fig. 5.6(a) and (b). The three-quarter wavelength line is composed of the phase inverter and two 70-Ω 45° synthesized CPWs. The remaining part of the coupler are replaced by 70-Ω 90° synthesized CPWs. The chip size is 1.915×1.625 mm^2, and the center operating frequency is 10 GHz. When compared with a conventional 1.5-λ_g ring coupler, the size reduction ratio reaches up to 94.7%. With the sum port being excited, the simulated and measured *S*-parameters and the phase differences between the outputs are summarized in Fig. 5.7. At the center frequency (10 GHz), the measured reflection coefficient $|S_{11}|$ and isolation $|S_{41}|$ are −25.7 and −20 dB,

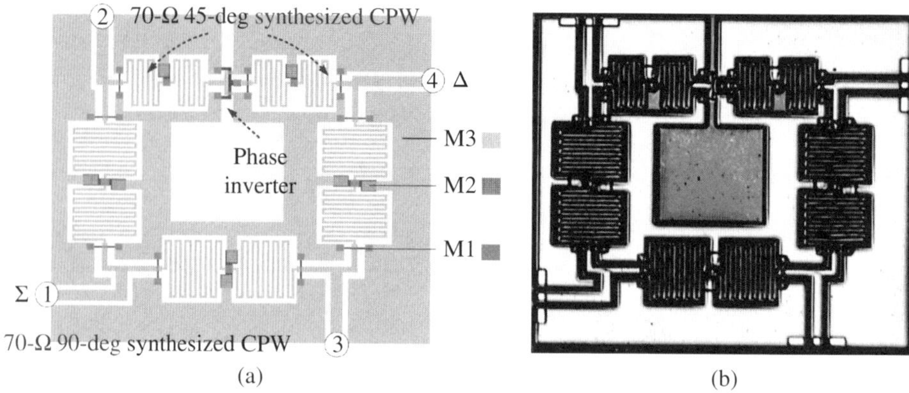

Figure 5.6 (a) Circuit layout and (b) chip photo of the miniaturized wideband rat-race coupler

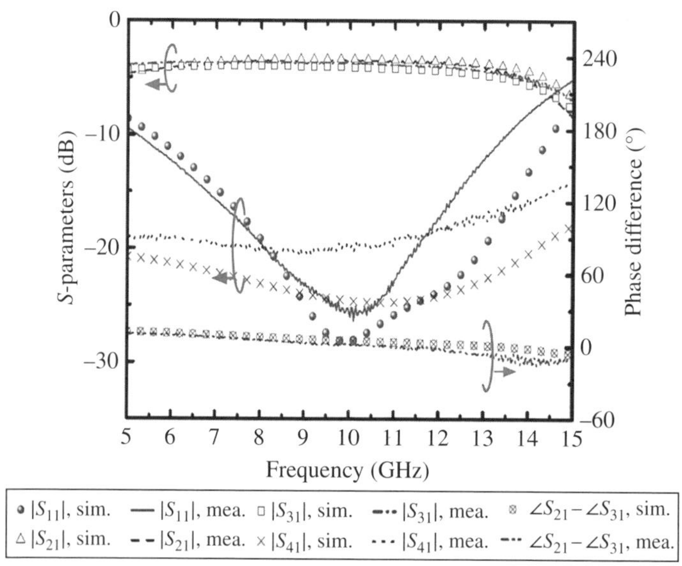

Figure 5.7 *S*-parameters of the miniaturized wideband rat-race coupler when the sum port is excited

respectively. The transmission coefficients $|S_{21}|$ and $|S_{31}|$, meanwhile, are −3.5 and −3.8 dB. The measured phase difference between the output ports ($\angle S_{31}-\angle S_{21}$) is 2.5°. The on-chip rat-race coupler shows a wide operating bandwidth from 6.8–12.4 GHz, or equivalently, a fractional bandwidth of 55%.

5.3.3 Dual-Band Rat-Race Coupler

A dual-band rat-race coupler can be realized by a proper connection of transmission-line sections loaded by shunt reactive elements [19]. A typical transmission-line network for dual-band operation is illustrated in Fig. 5.8. It is composed of a uniform line of (Z_1, 2θ) and two

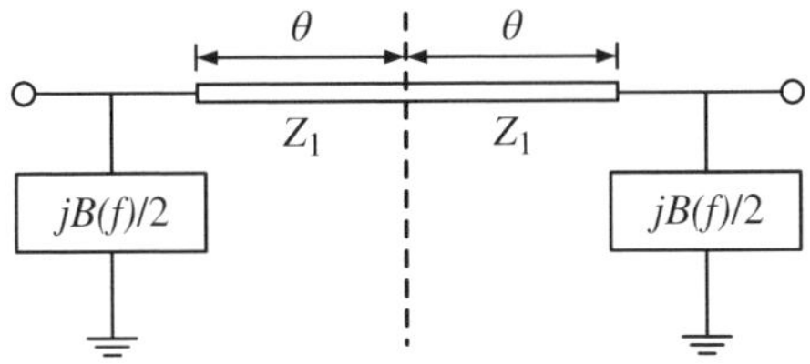

Figure 5.8 Transmission-line network as a dual-band quarter-wavelength line

shunt reactive elements. The loadings are in parallel with the line and placed at the input/output ports. By equating the Y matrix of the transmission-line network in Fig. 5.8 to that of a uniform Z_T-Ω, quarter-wavelength line, one gets

$$\begin{bmatrix} Y_{11} & Y_{21} \\ Y_{12} & Y_{22} \end{bmatrix} = \begin{bmatrix} \frac{j}{2}\left[B(f)+\frac{\tan\theta-\cot\theta}{Z_1}\right] & \frac{j}{2Z_1}(\tan\theta+\cot\theta) \\ \frac{j}{2Z_1}(\tan\theta+\cot\theta) & \frac{j}{2}\left[B(f)+\frac{\tan\theta-\cot\theta}{Z_1}\right] \end{bmatrix} = \begin{bmatrix} 0 & j\frac{1}{Z_T} \\ j\frac{1}{Z_T} & 0 \end{bmatrix}. \tag{5.5}$$

$B(f)$ is the susceptance of the reactive loading.

Since the transmission-line network is symmetric with respect to the central plane, the even-odd mode analysis can be applied. The entries in the Y matrix, Y_{11} and Y_{21}, are represented by the even- and odd-mode input admittances as

$$Y_{11} = \frac{Y_{ine}+Y_{ino}}{2} = j\frac{B}{2} + j\frac{\tan\theta}{2Z_1} - j\frac{\cot\theta}{2Z_1} \tag{5.6}$$

$$Y_{21} = \frac{Y_{ine}-Y_{ino}}{2} = j\frac{\tan\theta}{2Z_1} + j\frac{\cot\theta}{2Z_1}. \tag{5.7}$$

Let f_1 and f_2 be the operating center frequencies of two bands and by setting $k=f_2/f_1$ ($f_1<f_2$), the following conditions should be satisfied according to (5.5)

$$Y_{11}(f_1) = Y_{11}(kf_1) = 0 \tag{5.8}$$

$$Y_{21}(f_1) = Y_{21}(kf_1) = j\frac{1}{Z_T}. \tag{5.9}$$

Substitution of (5.6)–(5.7) to (5.8)–(5.9) yields

$$\theta = \frac{1}{2}\left(\frac{\pi}{1+k}\right) \tag{5.10}$$

$$Z_1 = Z_T \csc\frac{\pi}{1+k} \tag{5.11}$$

$$B(f) = \frac{2}{Z_T}\cos\frac{\pi}{1+k} \tag{5.12}$$

$$B(kf) = -\frac{2}{Z_T}\cos\frac{\pi}{1+k}. \tag{5.13}$$

With a given k, design variables of the transmission-line network in Fig. 5.8 can be calculated using (5.10)–(5.13). For example, with $f_1=2.45$ GHz and $f_2=5.25$ GHz ($k=2.14$), one gets $Z_1=84\ \Omega$, $\theta=28.6°$, $B(f_1)=0.0153$ S, and $B(f_2)=-0.0153$ S for realizing a dual-band 70-Ω, 90° line in the operating bands. Note that the required susceptance is positive at the low-band frequency f_1, but becomes negative with the same magnitude at the high-band frequency f_2. This suggests that with a proper design, a simple series *LC* resonator can be used as the required reactive element. The input susceptance of a lossless series *LC* tank (L_R, C_R) is given by

$$jB(f) = \frac{1}{\dfrac{1}{j2\pi fC_R} + j2\pi fL_R}. \tag{5.14}$$

By substituting (5.14) into (5.12) and (5.13) and after manipulation, the design variables are solved as

$$C_R = B(f_1)\frac{\omega_2-\omega_1}{\omega_1\omega_2}. \tag{5.15}$$

$$L_R = \frac{1}{(\omega_2-\omega_1)B(f_1)}. \tag{5.16}$$

With a simple synthesis, the circuit layout of the series resonator is shown in Fig. 5.9.

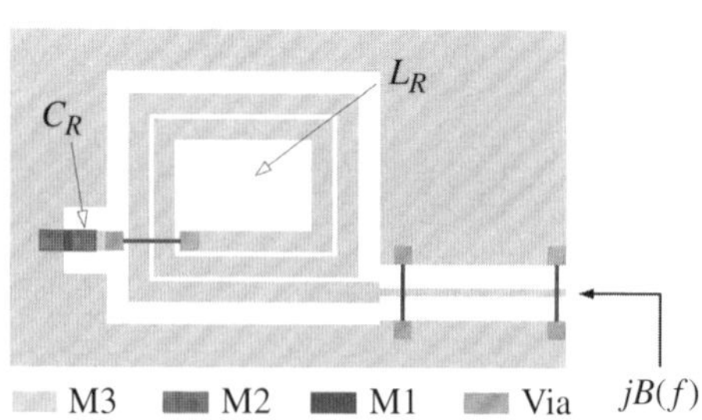

Figure 5.9 Layout of the series *LC* resonator for realizing the shunt loading

For on-chip realization, each uniform line section (Z_1, 2θ) is then replaced by a syntheszied CPW; the schematic diagram is shown in Fig. 5.10(a). The synthesized CPW is composed of five elements in a ladder-type connection. All inductors are set to the same value and so are the capacitors. By equating the five-element ladder-type network to a uniform transmission line of (Z_1, 2θ), the design equations are summarized as

$$\cos 2\theta = 1 - 3\omega^2 LC + \omega^4 L^2 C^2 \tag{5.17}$$

$$\sin 2\theta = \frac{2\omega L - \omega^3 L^2 C}{Z_1}. \tag{5.18}$$

The lumped parameters L and C can be solved using (5.17) and (5.18). The final layout of the synthesized CPW is plotted in Fig. 5.10(b).

By a direct integration of the series resonators and synthesized CPWs, the on-chip dual-band rat-race coupler is realized. The circuit layout is shown in Fig. 5.11; a chip photo is given as the inset. The design is simply completed by four elements, L_R, C_R, L, and C. The circuit size, excluding the pads, is 6.2×2.0 mm^2 or 0.00215 λ_g^2. When compared with a conventional single-band ring coupler (area=0.179 λ_g^2), the size reduction ratio is up to 98.8%.

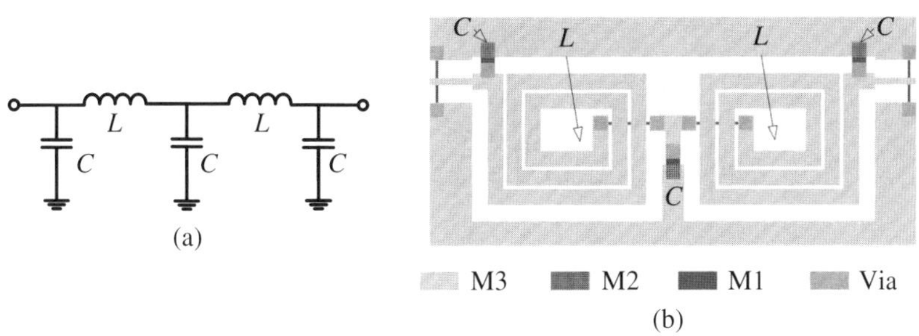

Figure 5.10 (a) Schematic diagram and (b) circuit layout of the synthesized CPW using five-element ladder-type network as a uniform line section (Z_1, 2θ)

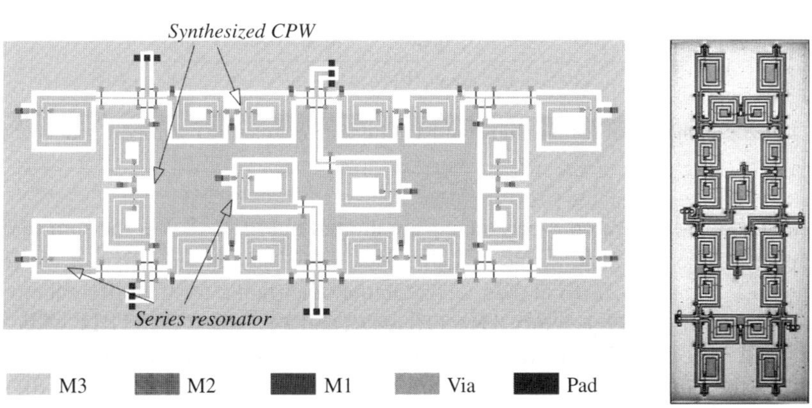

Figure 5.11 Layout and chip photo of the on-chip dual-band rat-race coupler using synthesized CPWs

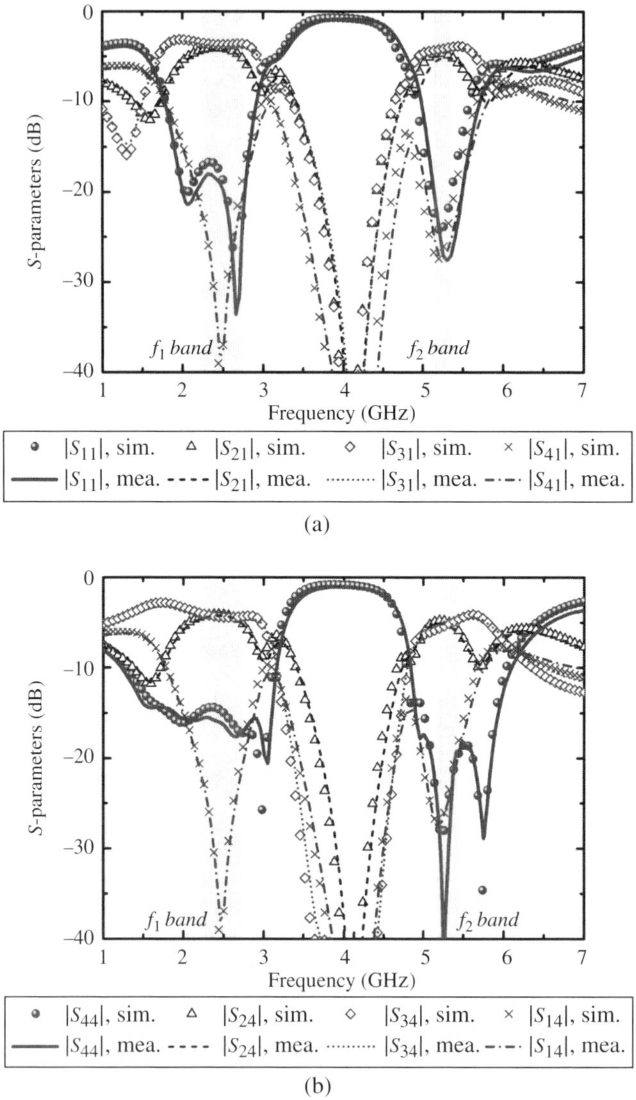

Figure 5.12 *S*-parameters of the on-chip dual-band rat-race coupler: (a) sum-port and (b) difference-port excitations

The simulated and measured *S*-parameters are plotted in Fig. 5.12. For the sum-port excitation, the measured $|S_{11}|$, $|S_{21}|$, $|S_{31}|$, and $|S_{41}|$ are −19, −3.9, −4.2, and −37 dB at 2.45 GHz; they are −27, −4.6, −5.1, and −27 dB at 5.25 GHz. On the other hand, for the difference-port excitation, the measured $|S_{44}|$, $|S_{34}|$, and $|S_{24}|$ are −16, −4.3, and −4.7 dB at 2.45 GHz, while they are −41, −5.1, and −5.9 dB at 5.25 GHz. The amplitude and phase imbalances are summarized in Fig. 5.13. For the sum-port excitation, the amplitude and phase imbalances are 0.4 dB and 0.8° at 2.45 GHz, and 0.4 dB and 0.5° at 5.25 GHz. For the difference-port excitation, the amplitude and phase imbalances are 0.5 dB and 176.7° at 2.45 GHz, and become as worse as 1.0 dB and 180.8° at 5.25 GHz.

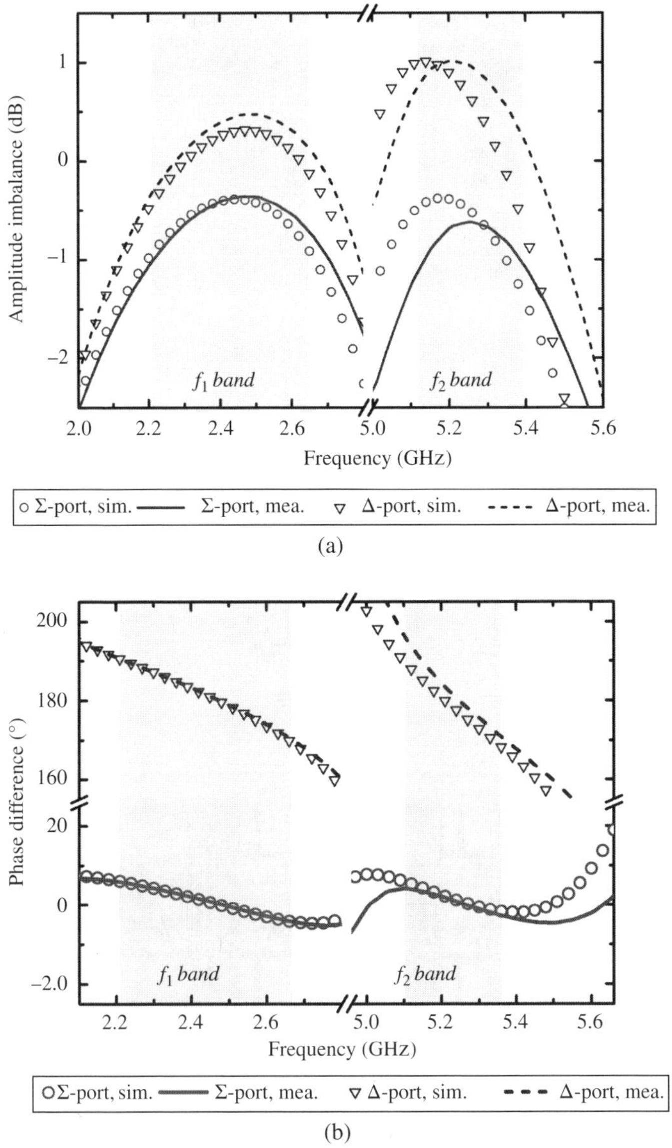

Figure 5.13 (a) Amplitude and (b) phase imbalances of the on-chip dual-band rat-race coupler

5.3.4 Coupled-Line Coupler

By introducing a tight coupling between two synthesized CPWs, an on-chip 3-dB backward-wave directional coupler is introduced in this section [6]. The coupler is developed based on the same principle discussed in Sec. 2.4.2, except that three sections of coupled synthesized CPWs are incorporated to complete the design. Figure 5.14(a) and (b) shows the circuit layout and equivalent lumped model of the on-chip coupler. The nominal center operating frequency is 4 GHz. The coupled-line coupler comprises three coupled synthesized CPWs in cascade.

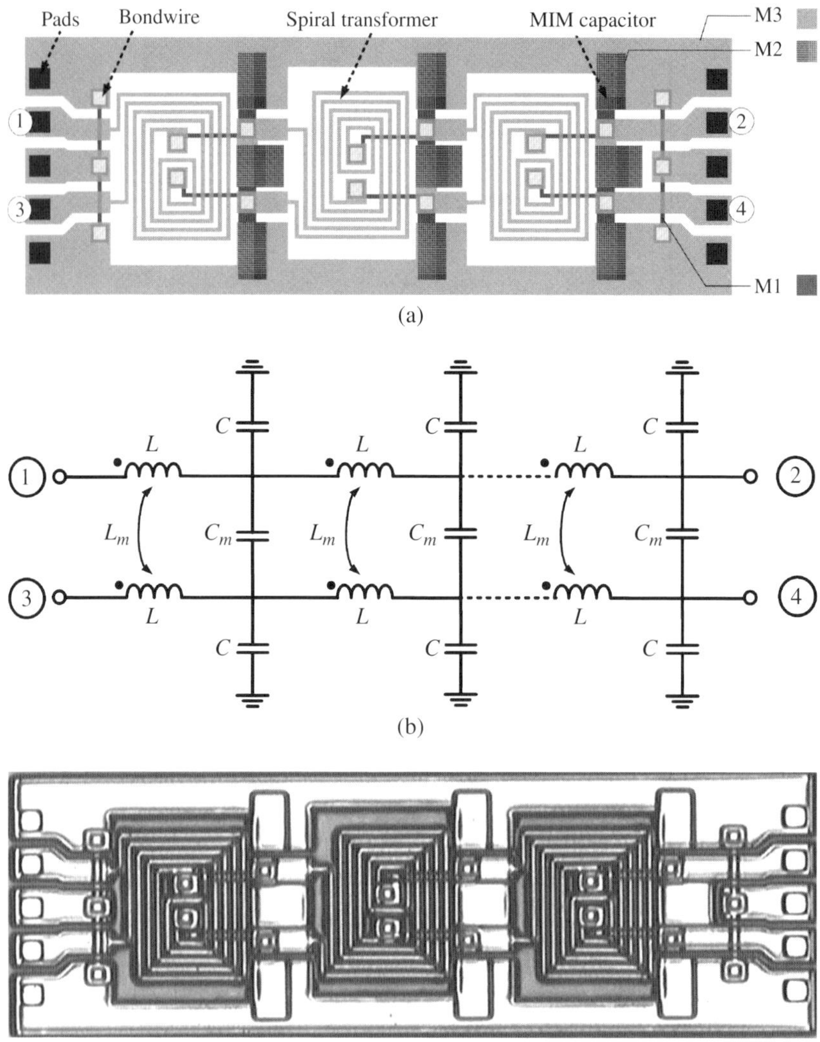

Figure 5.14 (a) Circuit layout, (b) equivalent lumped model, and (c) chip photo of the on-chip 3-dB coupled-line coupler. Source: Tseng 2014 [6]. Reproduced with permission of IEEE

Each section consists of a series inductor L_s (M3) and a shunt-to-ground capacitor C_p (M2/M3). The magnetic fields of the synthesized CPWs are coupled to each other by the mutually wound inductors in spiral form. It can be viewed as a spiral transformer and the mutual inductance is represented by L_m. The electrical coupling between the lines is achieved by the mutual capacitor C_m using MIM capacitors between the M2/M3 layers. A chip photo of the coupler is given in Fig. 5.14(c).

Referring again to Fig. 5.14(a), due to practical layout restrictions, the two spiral inductors in the transformer actually differ in electrical length by a small amount. To compensate for this phase difference, the spiral transformer in the middle stage is reversed in configuration with respect to the other two in the first and third sections. It helps equalize the overall

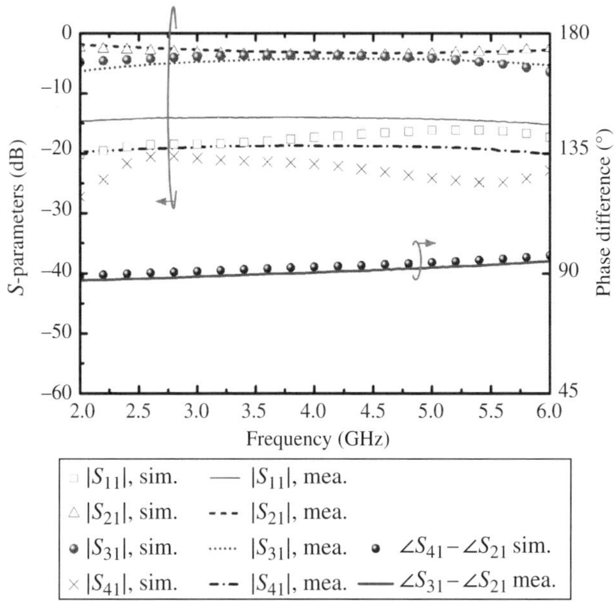

Figure 5.15 *S*-parameters and output phase differences of the on-chip coupled-line coupler

electrical lengths of the two synthesized CPWs and hence improves the directivity of the coupled-line coupler.

The simulated and measured *S*-parameters are shown in Fig. 5.15; the phase differences between the thru and coupled ports are illustrated at the same time. At the frequency with maximum coupling (4.32 GHz), the reflection coefficient ($|S_{11}|$) and isolation ($|S_{41}|$) are −14.1 and −18.7 dB, while the transmission coefficient ($|S_{21}|$) and coupling coefficient ($|S_{31}|$) are −3.3 and −4.2 dB, respectively. The measured phase difference is −90.8°. The frequency shift is explained by the process variation, specifically the thickness variation between the M2 and M3 layers for implementing the mutual capacitors C_m in this design. It makes the mutual capacitance lower than the nominal value, hence decreasing the coupling strength and moving the center frequency to a higher one. The chip size, excluding the pads, is 1.364×0.52 mm^2. When compared with a traditional design, the reduction in size is 73% while the reduction in coupling length reaches up to 85%.

5.3.5 Butler Matrix

As mentioned in Chapter 2, a Butler matrix consists of quadrature hybrids, crossovers, and pairs of delay lines. By replacing the line segments with synthesized transmission lines on the IPD substrate, an on-chip 4×4 Butler matrix is demonstrated in [8]. Referring to Fig. 5.16, the key building block is the bridged T-coil, a sort of non-periodic synthesized line with extraordinary size miniaturization capability [20, 21]. The bridged T-coil comprises a T-network in parallel with a capacitor C_p. The T-network is right-handed with two series inductors L_s separated by a shunt capacitor C_s. The pair of inductors is magnetically coupled to one another and the mutual inductance is L_M.

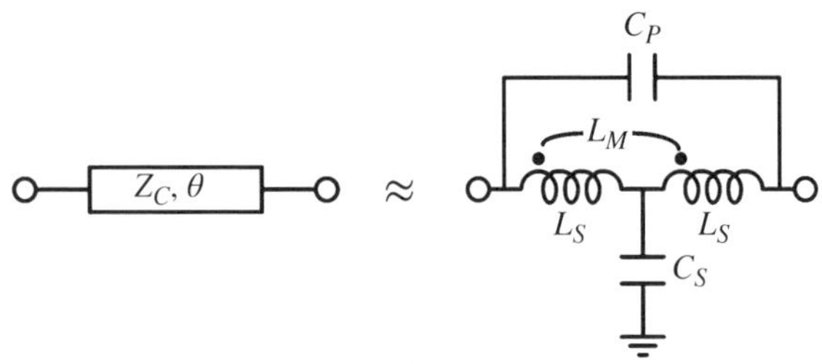

Figure 5.16 Bridged T-coil and its transmission line equivalence. Source: Lin 2013 [8]. Reproduced with permission of IEEE

The design parameters of the bridged T-coil can be achieved by equating its Z- and Y-matrices to those of a uniform transmission line of (Z_C, θ). To illustrate the synthesis procedure, the two-port Z- and Y-parameters of the bridged T-coil are first derived as

$$Y_{11} = Y_{22} = \frac{j\omega L_S + (j\omega C_S)^{-1}}{-\omega^2 (L_S^2 - L_M^2) + \frac{2(L_S + L_M)}{C_S}} + j\omega C_P \tag{5.19}$$

$$Y_{12} = Y_{21} = \frac{j\omega L_M - (j\omega C_S)^{-1}}{-\omega^2 (L_S^2 - L_M^2) + \frac{2(L_S + L_M)}{C_S}} - j\omega C_P \tag{5.20}$$

$$Z_{11} = Z_{22} = \frac{j\omega L_S + (j\omega C_S)^{-1} + j\omega C_P \left[-\omega^2 (L_S^2 - L_M^2) + \frac{2(L_S + L_M)}{C_S} \right]}{1 - 2\omega^2 C_P (L_S + L_M)} \tag{5.21}$$

$$Z_{12} = Z_{21} = \frac{-j\omega L_M + (j\omega C_S)^{-1} + j\omega C_P \left[-\omega^2 (L_S^2 - L_M^2) + \frac{2(L_S + L_M)}{C_S} \right]}{1 - 2\omega^2 C_P (L_S + L_M)}. \tag{5.22}$$

From (5.19) and (5.20), one gets

$$Y_{11} + Y_{12} = \frac{j\omega}{-\omega^2 (L_S - L_M) + \frac{2}{C_S}}. \tag{5.23}$$

For given values of $Y_{11a,b}$ and $Y_{12a,b}$ at two frequencies, say, f_a and f_b, the difference between L_S and L_M can be solved as

$$L_S - L_M = \frac{1}{\omega_b^2 - \omega_a^2} \left(\frac{j\omega_a}{Y_{11a} + Y_{12a}} - \frac{j\omega_b}{Y_{11b} + Y_{12b}} \right). \tag{5.24}$$

Substitution (5.24) into (5.23) yields

$$C_S = \frac{2}{\dfrac{j\omega_a}{Y_{11a}+Y_{12a}} + \omega_a^2\left(\dfrac{1}{\omega_b^2-\omega_a^2}\left(\dfrac{j\omega_a}{Y_{11a}+Y_{12a}} - \dfrac{j\omega_b}{Y_{11b}+Y_{12b}}\right)\right)}. \tag{5.25}$$

Likewise, by using the *Z*-parameters listed in (5.21) and (5.22) at the same frequencies, the sum of the L_S and L_M, as well as the parallel capacitor C_p can be solved as,

$$C_P = \frac{\dfrac{j\omega_a}{Z_{11a}-Z_{12a}} - \dfrac{j\omega_b}{Z_{11b}-Z_{12b}}}{2\left(\omega_b^2-\omega_a^2\right)} \tag{5.26}$$

$$L_S + L_M = \left(\frac{j\omega_a}{Z_{11a}-Z_{12a}} + 2\omega_a^2 C_p\right)^{-1}. \tag{5.27}$$

For a given set (Y_{11a}, Y_{12a}, Z_{11a}, Z_{12a}, Y_{11b}, Y_{12b}, Z_{11b}, Z_{12b}) of a uniform transmission line at the targeted operating frequencies f_a and f_b, the design variables (L_S, L_M, C_S, C_P) of the bridged T-coil can be solved from (5.24)–(5.27). The lumped values are then converted to physical footprints and integrated to complete the synthesized line.

Figure 5.17(a) shows the layout of the quadrature hybrid realized by pairs of bridged T-coils respectively having characteristic impedances of 35 and 50 Ω. Typical transmission responses are shown in Fig. 5.17(b). The center frequency is 2.5 GHz, and the chip size is 1.47×1.53 mm^2. The insertion loss is 4.5 ± 0.4 dB, and the operating bandwidth is 18%. The

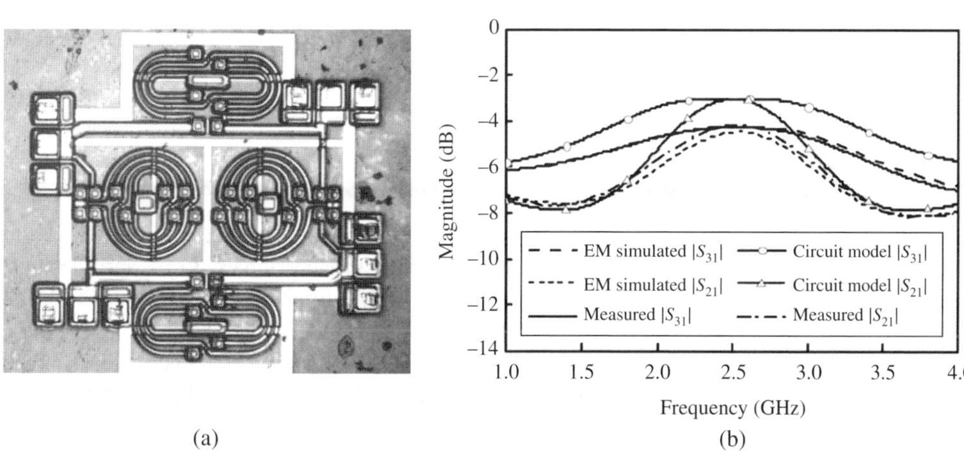

Figure 5.17 (a) Chip photo and (b) transmission coefficients of the quadrature hybrid by bridged T-coils. Source: Lin 2013 [8]. Reproduced with permission of IEEE

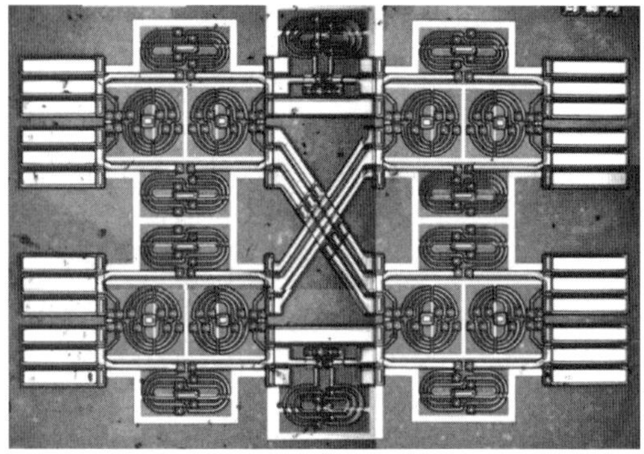

(a)

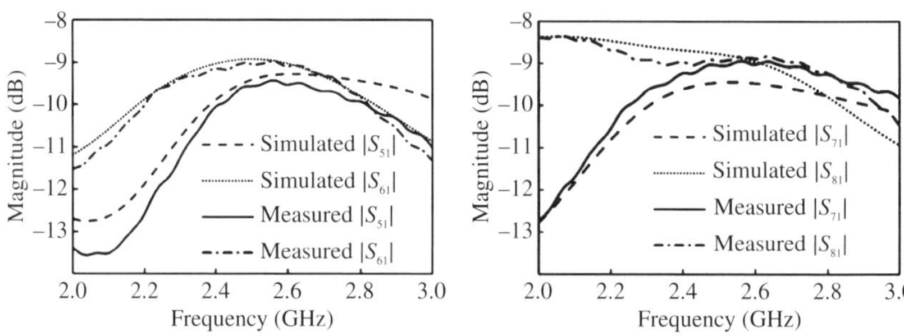

(b)

Figure 5.18 (a) Chip photo and (b) transmission coefficients of the on-chip Butler matrix by bridged T-coils. Source: Lin 2013 [8]. Reproduced with permission of IEEE

chip photo of the Butler matrix, by integrating the quadrature hybrids, is shown in Fig. 5.18(a). The chip size is $3.13 \times 3.3\ \mathrm{mm}^2$. The crossover coupler is realized by crossing the signal traces on the M1 and M3 layers. The transmission responses, by mounting the Butler matrix chip on a Rogers RO4350B substrate, are shown in Fig. 5.18(b). The extra power loss of the network, in average, is 3–4 dB.

5.4 Bandpass/Bandstop Filters Using Synthesized CPWs

Bandpass and bandstop filters (BPFs/BSFs) are indispensable components in both wired and wireless transceiver modules for rejecting unwanted interference signals. In modern designs, they are commonly realized by surface acoustic wave (SAW) filters and ceramic filters. While the SAW filters suffer from high insertion loss, the ceramic ones, such as the LTCC filters, require additional fabrication processes for integration. As an emerging technology in the system-in-package integration, the IPD process, especially the silicon-based IPD, finds advantages over other solutions in the implementation of passive components. Following this trend, a number of works devoted to developing bandpass/bandstop filters on the IPD substrate are reported [9–13]. Typical examples, developed with synthesized stepped-impedance resonators, transformer-coupled networks, and bridged T-coils are described in the following subsections.

5.4.1 Bandpass Filter Using Synthesized Stepped-Impedance Resonators

The stepped-impedance resonator (SIR), a proven technique to provide extended stopband, reduced footprint, and multiband operation, has been widely used in microwave filter designs. In this subsection, an on-chip three-pole SIR bandpass filter on the IPD process is introduced [11]. Each half-wavelength SIR in the chip design is realized by four synthesized CPWs in cascade, all comprising quasi-lumped inductors and capacitors. While the coupling between adjacent stages is feasible by high-Q MIM capacitors between the M1/M2 layers, the inductors are designed on the thick metal layer M3. Profiting from the slow-wave high/low-impedance synthesized CPWs, the on-chip filter shows an extraordinarily compact size with an extended stopband.

Figure 5.19(a) shows the general circuit model of a three-pole SIR BPF using capacitive coupling. Without changing the filter response, the coupling capacitance $C_{j,j+1}$ can be transformed into a standard J-inverter by adding pairs of fictitious shunt capacitances ($+C_{j,j+1}$, $-C_{j,j+1}$) to both sides of the capacitance as indicated in Fig. 5.19(b). The value of the J-inverter ($J_{j,j+1}$) is $\omega C_{j,j+1}$. The jth SIR resonator, accordingly, is loaded by two positive shunt capacitances ($C_{j-1,j}$, $C_{j,j+1}$) at the junction of the high/low-impedance lines. If the positive capacitances are comparable in value to that of the resonator, they serve as extra loading for size miniaturization. For simple analysis, hereafter the outermost negative capacitances are neglected as they can be easily compensated for by adjusting the I/O lines.

The schematic of the capacitively loaded SIR at the jth stage is shown in Fig. 5.20. The high/low-impedance lines have characteristic impedances and electric lengths of (Z_1,θ_1) and (Z_2,θ_2), respectively. Assuming that the coupling capacitances at adjacent stages are quite close to each other (i.e., $C_{j-1,j} = C_{j,j+1}$), the loaded SIR in Fig. 5.20 becomes approximately symmetric with respect to the central plane, and can be analyzed using the even-odd mode analysis. The formula for the fundamental (odd) mode, under the constraint of $\theta_1 = \theta_2 = \theta_m$, is simplified to,

$$\tan^2\theta_m + Z_2\omega_0 C_{j,j+1}\tan\theta_m - R = 0. \tag{5.28}$$

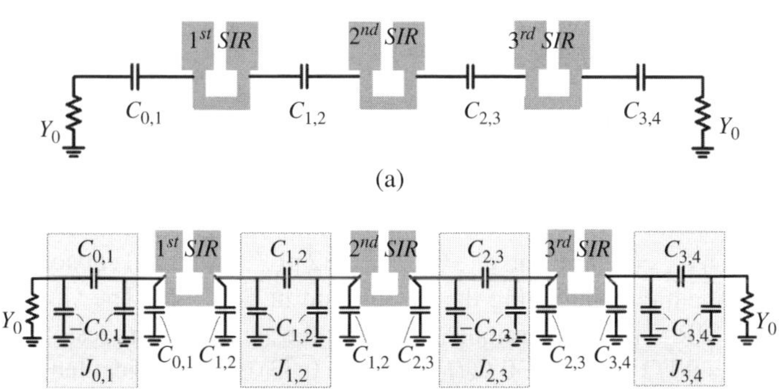

Figure 5.19 (a) General scheme and (b) equivalent model with J-inverters of a three-pole SIR BPF using capacitive coupling. Source: Tseng 2014 [11]. Reproduced with permission of IEEE

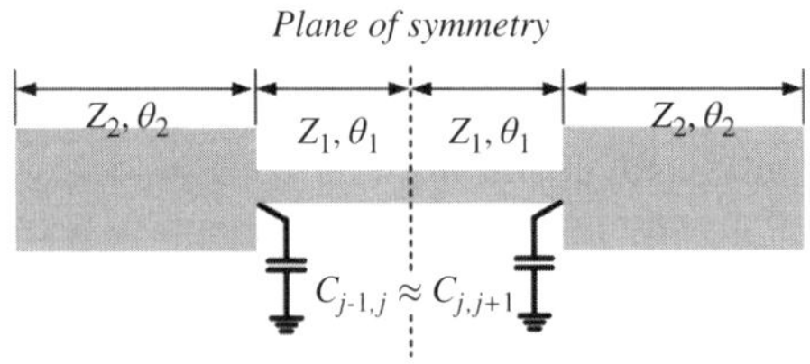

Figure 5.20 Schematic of the capacitively loaded SIR at the *j*th stage. Source: Tseng 2014 [11]. Reproduced with permission of IEEE

It is a quadratic function of tan θ_m, with $R = Z_2/Z_1$ being the impedance ratio. Using (5.28), the design parameter θ_m can be solved uniquely in terms of Z_1, Z_2, and $C_{j,j+1}$. The next higher-order mode of the loaded SIR, occurring when $\theta_m = \pi/2$, is the first even mode; it can be solved in a similar manner.

Meanwhile, in accordance with the filter specification, the *J*-inverters are determined using the standard formulae [22];

$$J_{0,1} = \sqrt{\frac{Y_o b_1 FBW}{g_0 g_1}} \tag{5.29a}$$

$$J_{1,2} = FBW\sqrt{\frac{b_1 b_2}{g_1 g_2}} \tag{5.29b}$$

$$J_{2,3} = \sqrt{\frac{Y_o b_3 FBW}{g_2 g_3}} \tag{5.29c}$$

$$b_i = \frac{\omega_o}{2}\frac{dB_i(\omega)}{d\omega}\bigg|_{\omega=\omega_o} \quad (i = 1, 2, 3). \tag{5.29d}$$

The mutual (coupling) capacitance is therefore:

$$C_{j,j+1} = J_{j,j+1} / \omega_0. \tag{5.30}$$

By substituting $C_{j,j+1}$ and R into (5.28), the desired electrical length of the high/low-impedance sections can be solved routinely.

In this specific design, a Chebyshev prototype with a 0.01-dB ripple is selected. With f_0 = 4 GHz, FBW = 30% and slope parameter $b_j = 0.054$, the coupling capacitances are calculated from (5.30) as $C_{0,1} = C_{3,4} = 0.9$ pF and $C_{1,2} = C_{2,3} = 0.82$ pF. The mutual capacitances at adjacent stages are almost identical, validating the assumption in Fig. 5.20. The slope parameter b_i is extracted from an SIR using conventional CPWs on the same IPD process.

By selecting $Z_1 = 90\ \Omega$ and $Z_2 = 30\ \Omega$ ($R = Z_2/Z_1 = 0.33$), the electrical length θ_m is solved as 18.7° if the average of $C_{0,1}$ and $C_{1,2}$ is used in (5.28). Clearly, the capacitively loaded λ/2 SIR pushes the fundamental odd mode further to the lower frequency side whereas the next even mode remains unaltered. When compared with a conventional SIR with the same ratio *R*, the one with strong mutual coupling has a shorter physical length and a wider bandgap between the fundamental and next higher-order modes (spurious passband).

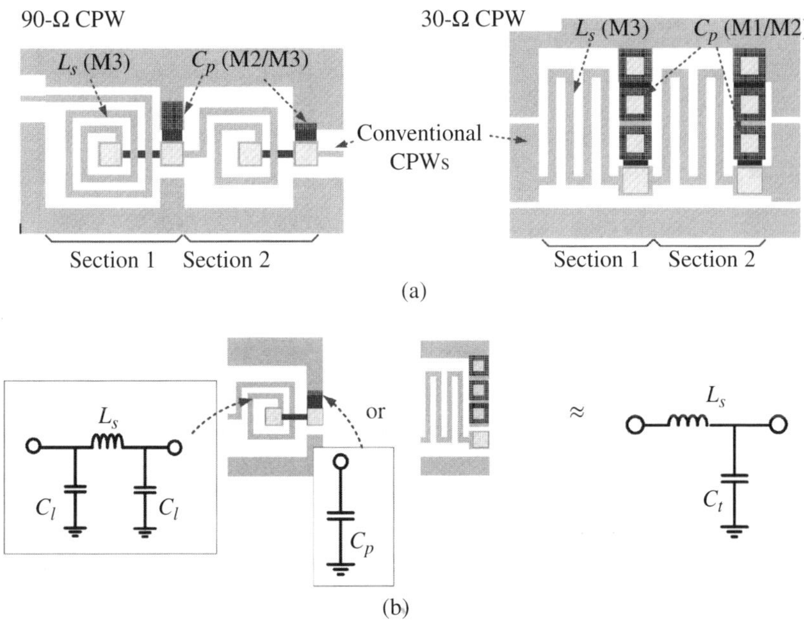

Figure 5.21 (a) Layouts and (b) approximate lumped model of the synthesized CPWs

The next step is to reconstruct all line segments of the SIR using synthesized CPWs with quasi-lumped elements on the IPD process. The layouts of the 90-Ω and 30-Ω synthesized CPWs are depicted in Fig. 5.21(a). Each synthesized CPW is composed of two sections, with each associated with a series inductor (M3) and a shunt MIM capacitor (M1/M2 or M2/M3). For clarity, the lumped equivalent model is shown in Fig. 5.21(b). The synthesis procedure is similar to that in Chapter 2. Note that spiral inductors are used in the high-impedance line, while the low-impedance line is realized by meander-line inductors with lower inductance values.

Following the filter design procedure and SIR synthesis approach, the final layout and chip photo of the IPD three-pole BPF are shown in Fig. 5.22. To optimize the area usage, each synthesized SIR is folded such that the three stages are filled into a rectangular region perfectly. All coupling capacitors are fine-tuned as 0.92 pF to account for the slight variation of the performance of the synthesized CPWs. In terms of the guided wavelength, the filter has a very compact size of 1.948 × 1.266 mm^2 (0.045 × 0.029 λ_g^2).

The S-parameters of the SIR filter are illustrated in Fig. 5.23 from 1 to 26.5 GHz; the measured group delay in the passband is shown as an inset. The agreement between the results is fair. The discrepancy is mostly attributed to the fabrication variation of the substrate thickness. The first spurious passband, with a peak of $|S_{21}| \approx -10$ dB, is pushed to 16.7 GHz (~4.3f_o). At 4.2 GHz, the measured $|S_{11}|$ is −14.4 dB, while the $|S_{21}|$ is −3.2 dB. The operating bandwidth is 29% (3.5–4.7 GHz). The upper stopbands with $|S_{21}| < -20$ dB are from 5.9 to 15.8 GHz, and from 17.9 to more than 26.5 GHz. Three transmission zeros are observed at 8, 15.4, and 18.8 GHz. While the first transmission zero is accounted for by the parasitic cross-coupling, the remaining two at 15.4 and 18.8 GHz can be explained by the virtual short circuit of the low-impedance synthesized CPW with its electrical length equal to 90°. Profiting from the synthesized SIRs, the on-chip BPF features acceptable in-band power dissipation with a remarkable extended stopband.

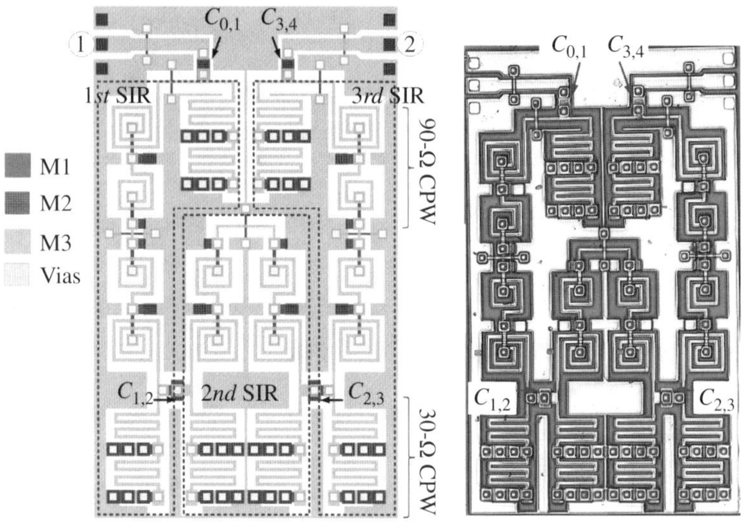

Figure 5.22 Layout and photograph of the SIR BPF. Source: Tseng 2014 [11]. Reproduced with permission of IEEE

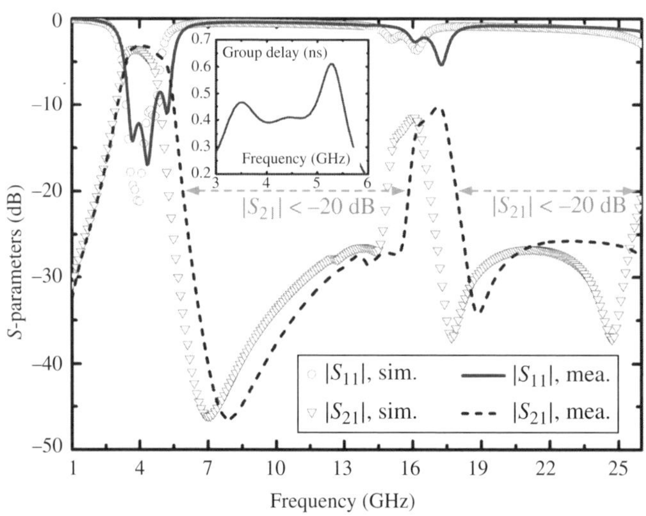

Figure 5.23 *S*-parameters and group delay of the SIR BPF. Source: Tseng 2014 [11]. Reproduced with permission of IEEE

5.4.2 Transformer-Coupled Bandpass Filter

The on-chip bandpass filter can be realized on the IPD substrate by a transformer [12], as well. Referring to the equivalent model in Fig. 5.24, each coil in the transformer is modeled by a π-network, consisting of a coil inductor (L_1–L_2–L_1) in the series path and two capacitors (C) in the shunt path. The π-model can be viewed as a non-periodic synthesized line using the

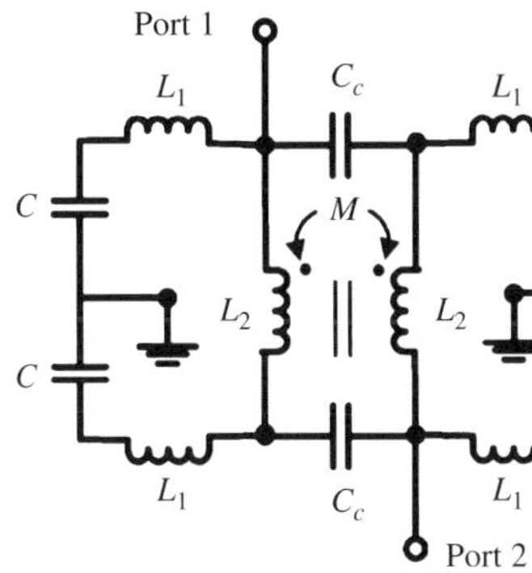

Figure 5.24 Equivalent model of the transformer-coupled BPF. Source: Chen 2010 [12]. Reproduced with permission of IEEE

three-element ladder-type network. The inductor and capacitor form a parallel resonator, which is coupled to the next stage using the mutual inductance M and mutual capacitance C_c.

The on-chip bandpass filter is synthesized using the standard coupled resonator theory with two design parameters: the coupling coefficient k and external quality factor Q_e. The specification of the bandpass filter is related to the design parameters by

$$k_{j,j+1} = \frac{\Delta}{g_j g_{j+1}} \tag{5.31a}$$

$$Q_e = \frac{g_0 g_1}{\Delta} = \frac{g_n g_{n+1}}{\Delta}. \tag{5.31b}$$

$k_{j,j+1}$ is the coupling coefficient between the jth and $(j+1)$th resonators; Δ is the fractional bandwidth and g_j is the jth prototype element value. The interested reader may find synthesis details in [22, 23].

Figure 5.25(a) shows the chip photo of the on-chip design; the size is less than 0.7×1.1 mm^2. It is a two-pole filter based on a 0.01-dB equal-ripple Chebyshev response. The fractional bandwidth is 15%. The transformer is in octagonal shape with the parallel coils symmetrically wound side-by-side on M3. The capacitors are attached to the ends of the coils. Figure 5.25(b) plots the S-parameters of the BPF. The passband is from 2.7 to 2.9 GHz with an insertion loss of 2.2 dB. Two transmission zeros are observed at 1 and 4 GHz. The higher transmission zero mainly results from the series resonance of L_1 and C, while the lower one can be accounted for by the cross-coupling effect.

5.4.3 *Bridged T-Coils as Common-Mode Filter*

Differential signals provide high immunity to ambient noise and can significantly reduce deterioration from nonideal signal return paths. They have been widely implemented in high-speed data transmissions. In reality, however, due to various restrictions in the layout assignment, bends in the routing paths and mismatch of the signal transmission and return paths are

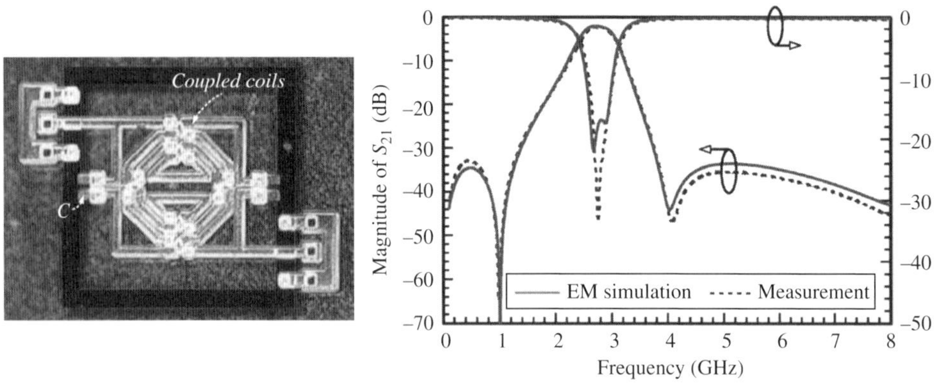

(a) (b)

Figure 5.25 (a) Chip photo and (b) *S*-parameters of the transformer-coupled BPF. Source: Chen 2010 [12]. Reproduced with permission of IEEE

inevitable, which may convert the differential signals into unwanted common-mode noises and electromagnetic leakages. To alleviate the degradation, the common-mode rejection filter is a key component in high-speed differential systems to suppress the unwanted electromagnetic interference (EMI) and preserve signal integrity.

The bridged T-coil, or the modified-T circuit, can be applied to realize an on-chip common-mode filter with high performances [13]. Figure 5.26(a) shows the unit cell of the bridged T-coils implemented in the on-chip common-mode rejection filter. It actually consists of two bridged T-coils jointed together at the point *P*; an extra inductor L_v shunted to ground is connected to this junction to complete the design. Since the unit cell is symmetric with respect to the central horizontal plane, the even-odd mode analysis can be applied, and the half circuits are summarized in Fig. 5.26(b).

Referring to Fig. 5.26(b), in the odd mode, the half circuit is no more than a routine bridged T-coil. Its electrical properties can be characterized as a uniform transmission line of (Z_c, θ) at an angular frequency ω using the following equations,

$$L_s = Z_c \frac{\theta}{\omega}\left(0.25 + \frac{1}{\pi^2}\right) \tag{5.32}$$

$$C_s = \frac{\theta}{\omega Z_c} \tag{5.33}$$

$$L_m = Z_c \frac{\theta}{\omega}\left(0.25 - \frac{1}{\pi^2}\right) \tag{5.34}$$

$$C_m = \frac{\theta}{\omega \pi^2 Z_c}. \tag{5.35}$$

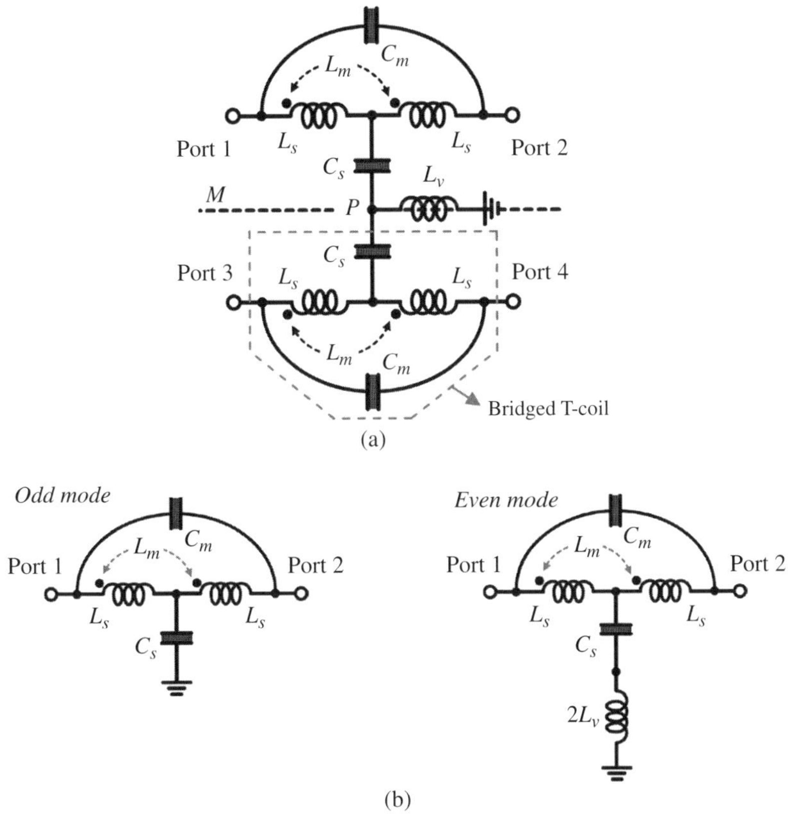

Figure 5.26 (a) Unit cell and (b) even/odd-mode half circuits of the bridged T-coil for realizing the common-mode filter. Source: Hsiao 2015 [13]. Reproduced with permission of IEEE

In the even mode, however, the two-port network becomes a bandstop filter with two transmission zeros. The transmission zeros are contributed not only by the series resonator ($2L_v$, C_s) but also the upper part of the network. Since the two-port even-mode network is also symmetrical with respect to the central vertical plane, its transmission coefficient can be expressed by

$$S_{21} = \frac{Z_0\left(Z_{ine} - Z_{ino}\right)}{\left(Z_{ine} + Z_0\right)\left(Z_{ino} + Z_0\right)} \tag{5.36}$$

where

$$Z_{ine} = \frac{2 - \omega^2 C_s\left(L_s - L_m + 4L_v\right)}{j\omega C_s} \tag{5.37a}$$

$$Z_{ino} = \frac{j\omega\left(L_s + L_m\right)}{1 - 2\omega^2 C_m\left(L_s + L_m\right)}. \tag{5.37b}$$

By setting $Z_{ine} = Z_{ino}$, the frequencies of the transmission zeros with $S_{21} = 0$ can be determined from (5.36) and (5.37) as

$$\omega_{1,2} = \sqrt{\frac{-B \pm \sqrt{B^2 - 4A}}{2A}} \tag{5.38}$$

where

$$A = C_s C_m \left(L_s + L_m\right)\left(L_s - L_m + 4L_v\right) \tag{5.39a}$$

$$B = -2C_m \left(L_s + L_m\right) - C_s \left(2L_v - L_m\right). \tag{5.39b}$$

Substitution of (5.32)–(5.35) into (5.38) and after manipulation, the Propagation delay which is the electrical length θ over the angular operating frequency ω, is expressed in terms of the angular transmission-zero frequencies ω_1 and ω_2 as

$$\frac{\theta}{\omega} = \sqrt{\frac{\left(\omega_1^2 + \omega_2^2\right) - \sqrt{\left(\omega_1^2 - \omega_2^2\right)^2 + \left(\pi\omega_1\omega_2\right)^2}}{2\omega_1^2\omega_2^2\left(\frac{1}{\pi^2} - 0.25\right)}}. \tag{5.40}$$

Using (5.40), the equivalent electrical length of the bridged T-coil network operated in the odd mode can be determined with the pair of transmission zeros in the common mode being specified in advance. It is then substituted into (5.32)–(5.35) to solve the design parameters (L_s, L_m, C_s, C_m). The final variable, the series inductor L_v, is fixed at the same time using

$$L_v = \frac{Z_0}{2}\sqrt{\frac{\pi^2}{\omega_1^2\omega_2^2\left(\theta/\omega\right)^3} - \frac{\left(\theta/\omega\right)}{\pi^2}}. \tag{5.41}$$

By converting the design parameters into physical footprints, the layout of a single bridged T-coil using the IPD process is shown in Fig. 5.27(a). The inductors on the M1 and M3 layers are mirrored to each other to achieve the mutual inductance L_m. The mutual capacitance C_m, in the meantime, is realized by an interdigital capacitor on the M1 layer.

Three unit cells are then connected in cascade to fulfill the common-mode rejection filter. The chip photo is shown in Fig. 5.27(b); the size is only 1 × 0.95 mm^2. The series inductor L_v is actually realized by two inductors in parallel to keep the layout in full symmetry. Figure 5.28(a) and (b) shows the common-mode and differential-mode S-parameters of the filter, respectively. The common-mode rejection bandwidth with $|S_{cc21}| < -10$ dB is from 1.6–4.7 GHz, and the 3-dB cutoff frequency of the differential mode is about 10 GHz. The reflection coefficient in the differential mode $|S_{dd11}|$ is lower than −15 dB. The eye diagrams of 4 and 6 Gb/s, with 2-V PRBS (pseudorandom binary sequence) differential signals using a pattern generator, are shown in Fig. 5.28(c). The opening of the eye pattern is good under different data rates.

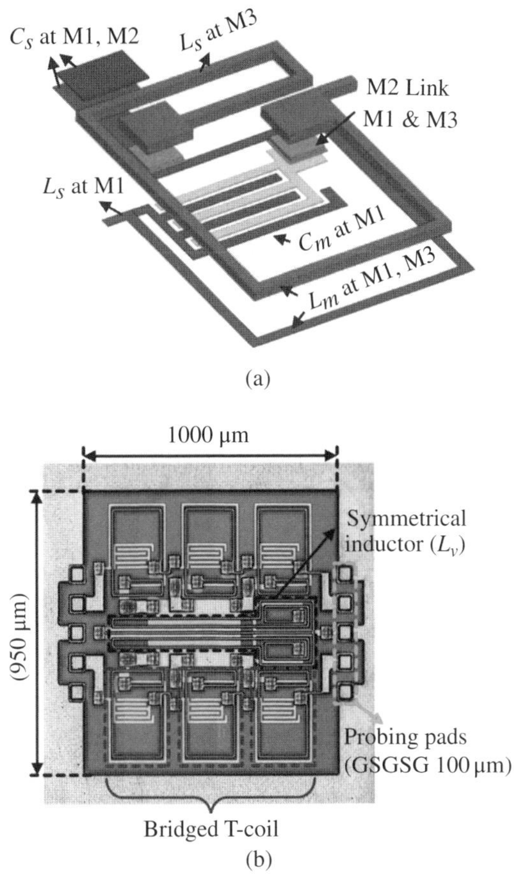

Figure 5.27 (a) Layout of the bridged T-coil and (b) chip photo of the common-mode filter on the IPD process. Source: Hsiao 2015 [13]. Reproduced with permission of IEEE

5.5 Chip Designs Using Multi-Mode Synthesized CPWs

The multi-operational mode synthesized lines studied intensively in Chapters 3 and 4 can be extended to fulfill on-chip designs, as well. Dual-mode or tri-mode synthesized CPWs are developed using the IPD process to realize a number of circuits not feasible with conventional components. In addition to on-chip multiplexers, the chip implementation of a dual-mode rat-race coupler for the decoupling network of a dual-band closely-spaced antenna array is demonstrated. An on-chip liquid detector for calculating the complex dielectric constant of liquids is developed with the aid of dual-mode synthesized CPWs as well.

5.5.1 Diplexer

The first example is an on-chip diplexer with an extraordinarily compact size [14]. The schematic diagram is the same as that in Sec. 3.5.1. The diplexer only comprises two essential building blocks. In the low band, the right-handed (RH) synthesized CPW is equivalent to a

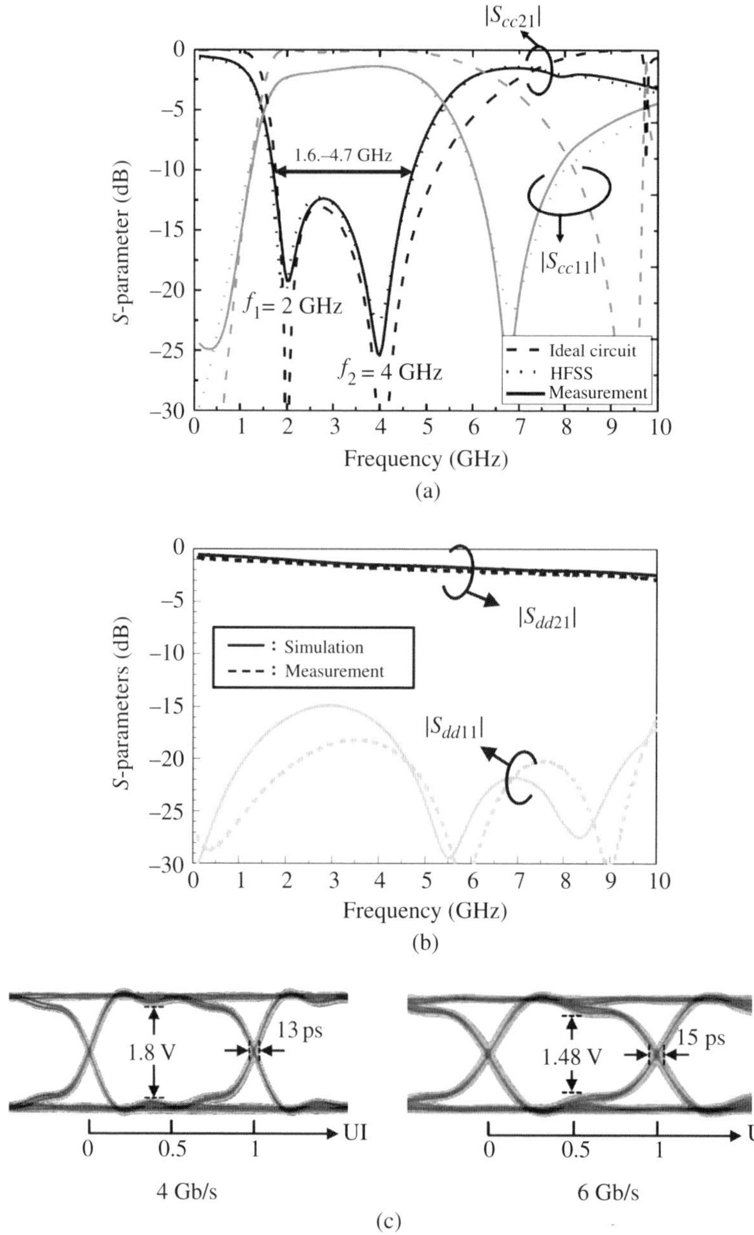

Figure 5.28 (a) Common-mode S-parameters, (b) differential-mode S-parameters, and (c) eye diagrams of the common-mode filter on the IPD process. Source: Hsiao 2015 [13]. Reproduced with permission of IEEE

uniform line and serves as the transmission path (Channel 1). The left-handed (LH) synthesized CPW, meanwhile, resembles an open circuit and makes the lower part invisible to the active channel. The operation in the high band is reversed, in which the RH synthesized CPW becomes open-circuited and the LH synthesized CPW sets up the transmission path (Channel 2).

The building blocks are realized by the dual-mode synthesized lines in Sec. 3.3.2. All quasi-lumped elements are in CPW form. Figure 5.29 shows the equivalent models and corresponding quasi-lumped layouts of the RH and LH dual-mode synthesized CPWs. A series *LC* tank ($L_{R2} + C_{R2}$ in the RH line and $L_{L2} + C_{L2}$ in the LH line) is placed at the plane of symmetry for introducing a transmission zero. The *LC* sections, with a shunt capacitor (C_{R1}) and series inductor (L_{R1}) in the RH line and a shunt inductor (L_{L1}) and series capacitor (C_{L1}) in the LH line, are connected to both sides of the resonator for fulfilling the impedance matching in the passband, as well as the open circuit condition at the frequency of transmission zero. The interested readers may refer to Sec. 3.3.2 for the detailed formulation.

The dual-mode RH and LH synthesized CPWs are directly integrated as the on-chip diplexer. Due to the open circuit condition in one of the lines, no extra matching unit is required at the common input. The layout and chip photo are shown in Fig. 5.30. The circuit area is only 1.125 × 1.105 mm^2 (0.0174 λ_g × 0.0171 λ_g). Figure 5.31 illustrates the simulated and measured *S*-parameters of the diplexer. In channel 1 (2.6 GHz), the measured $|S_{11}|$, $|S_{21}|$, $|S_{31}|$, and $|S_{32}|$ are −29, −0.8, −28, and −26 dB, while they are −23, −29, −0.6, and −39 dB in the

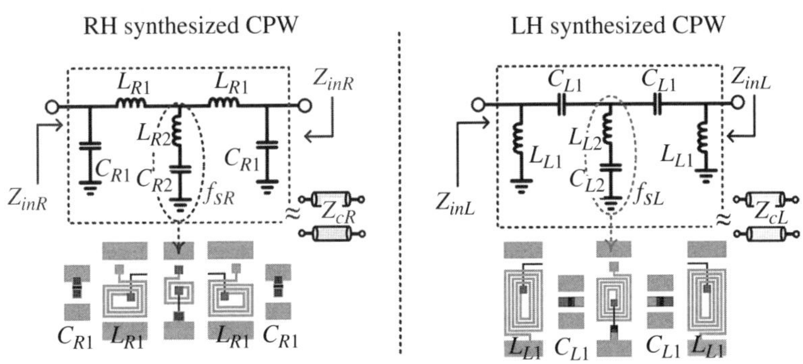

Figure 5.29 Equivalent circuit models and quasi-lumped layouts of the RH and LH synthesized CPWs. Source: Lai 2014 [14]. Reproduced with permission of IEEE

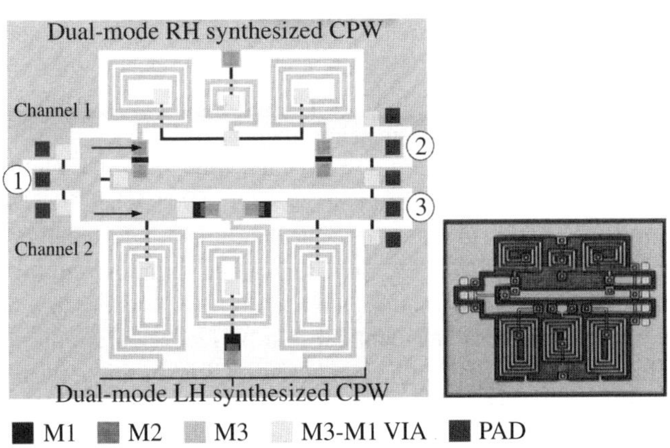

Figure 5.30 Layout and chip photo of the on-chip diplexer. Source: Lai 2014 [14]. Reproduced with permission of IEEE

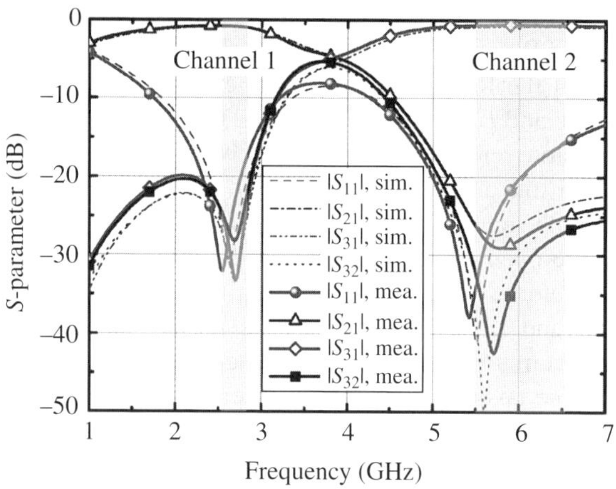

Figure 5.31 *S*-parameters of the on-chip diplexer. Source: Lai 2014 [14]. Reproduced with permission of IEEE

channel 2 (5.8 GHz). The fractional bandwidth is 10% in the channel 1 (2.54–2.82 GHz) and 18% in channel 2 (5.44–6.53 GHz). The transmission loss in the passbands is less than 1 dB.

5.5.2 Dual-Mode Rat-Race Coupler

The dual-mode rat-race coupler, an unusual component not familiar to the community, functions identical to its conventional counterpart in one band but auto-reconfigures as a pair of uncoupled transmission lines in the other band. In [24], it was applied to realize a dual-band decoupling network for a closely-spaced two-element antenna array. Although the performance of the decoupling network in the literature is good, its occupied size is too large to have practical applications. To tackle the problem, an on-chip version of the dual-mode rat-race coupler is demonstrated using the IPD process.

The schematic diagram of the dual-mode rat-race coupler is shown in Fig. 5.32(a). In the low band, the four-port network is identical to a conventional rat-race coupler with two orthogonal modes at the inputs (Σ and Δ ports). In the high band, however, the shunt paths between the upper and lower parts of the network are prohibited to enforce the four-port network identical to two isolated direct-thru transmission lines.

Figure 5.32(b) plots the circuit topology of the dual-mode coupler. Unlike a conventional design, it consists of a quadrature hybrid and an extra 50-Ω 90° delay line. Compared with the classical 1.5-λ_g ring coupler, this topology does not require an additional crossover for interconnection. The coupler is realized by two dual-mode synthesized lines with different responses. The dual-mode line in the shunt path should be equivalent to a 50-Ω 90° uniform transmission line in the low band, but identical to a perfect open circuit when looking into either side in the high band. Similar to the synthesized CPW in the previous subsection, this requirement is fulfilled by a five-element ladder-type network with a series resonator placed at the plane of symmetry. The chip photo and associated lumped model are shown in Fig. 5.33(a).

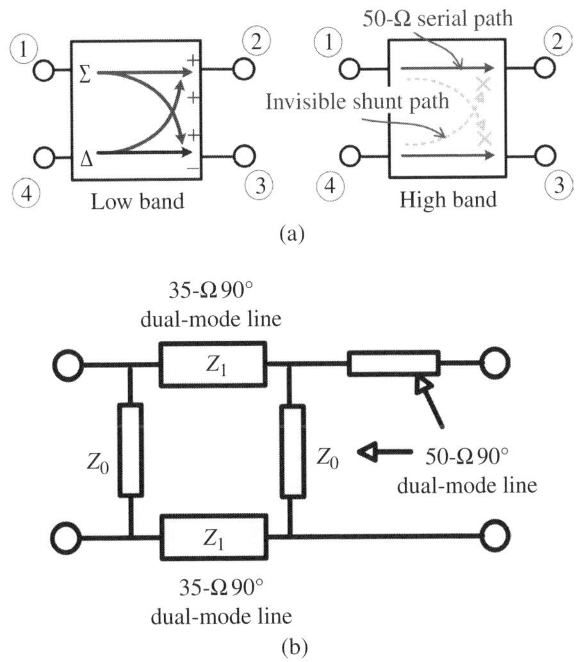

Figure 5.32 (a) Schematic diagram and (b) circuit topology of the dual-mode rat-race coupler

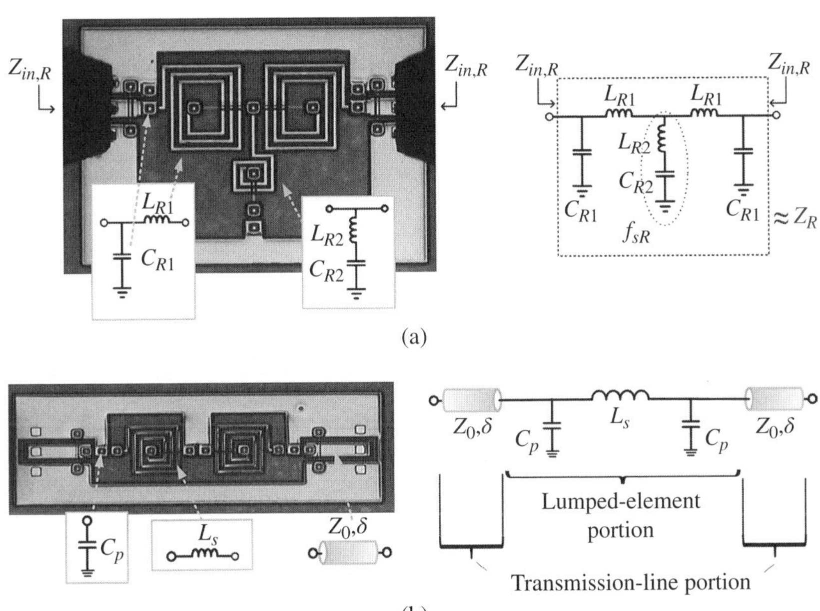

Figure 5.33 Equivalent lumped models and chip photos of the dual-mode synthesized CPWs in the (a) shunt path and (b) series path

The dual-mode synthesized line in the series path, in the meantime, plays a different scenario. It is identical to a 35-Ω 90° transmission line in the low band, but turns to a 50-Ω uniform line in the high band to avoid reflections from the antenna terminals and system inputs. This synthesized line is realized by the network in Fig. 5.33(b). The network comprises a lumped-element portion and a transmission-line portion. The lumped portion is a π-network with a series inductor L_s and two shunt capacitors C_p; the transmission-line portion, meanwhile, consists of two uniform 50-Ω transmission lines each associated with an electrical length of δ.

There are three requisite conditions the dual-mode line must comply with:

1. The characteristic impedance of the line must be 35 Ω at the center angular frequency of the low band (ω_L).
2. The electrical length is 90° at ω_L.
3. The characteristic impedance of the synthesized line must be 50 Ω at the center angular frequency of the high band (ω_H); at this frequency, the electrical length is of no concern.

Since the three-pole *LC* network and the transmission line sections are connected in cascade, the synthesized CPW can be analyzed using the *ABCD* matrices. In the low band, the entire network including the lumped-element and transmission-line portions should be identical to a 35-Ω 90° transmission line, while the lumped portion alone must be equivalent to a 50-Ω transmission line in the high band. After equating the *ABCD* matrix of the dual-mode network to that of a uniform transmission line with given electrical properties at both bands, a cubic equation of the shunt capacitance C_p can be formulated as

$$\alpha C_p^3 - \beta C_p^2 - |K| = 0, \tag{5.42}$$

where

$$\alpha = 2\omega_L Z_0^3\left(\omega_H^2 - \omega_L^2\right) \tag{5.43}$$

$$\beta = \omega_H^2 Z_0^2 |K| \tag{5.44}$$

and

$$K = {Z_1}/{Z_0} - {Z_0}/{Z_1}. \tag{5.45}$$

With the shunt capacitance explicitly solved, the series inductance L_s is given by

$$L_s = \frac{2Z_0^2 C_p}{1+\omega_H^2 Z_0^2 C_p^2}. \tag{5.46}$$

The electrical length of the transmission-line section can be determined at the same time,

$$\delta = \frac{1}{2}\cot^{-1}\sqrt{\frac{K^2(M+4)}{4\left(M-K^2\right)}} \tag{5.47}$$

$$M = \frac{\left(\omega_H^2 - \omega_L^2\right)^2 Z_0^4 C_p^4}{1+\left(\omega_H^2 - \omega_L^2\right) Z_0^2 C_p^2}. \tag{5.48}$$

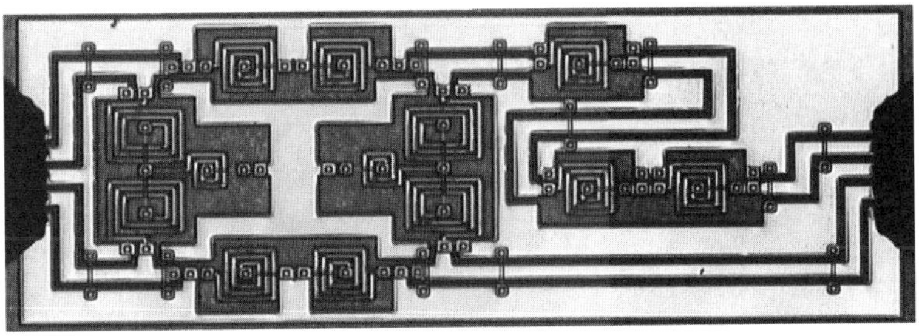

Figure 5.34 Chip photo of the dual-mode rat-race coupler

The interested readers may find more information regarding the formulation in [24]. With $Z_1 = 35\ \Omega$, $Z_0 = 50\ \Omega$ and the two operating frequencies specified, the variables (L_s, C_p, and δ) can be uniquely solved to complete the design. The chip photo of the dual-mode synthesized CPW to be implemented in the series path of the coupler is shown in Fig. 5.33(b).

By a direct integration of the dual-mode synthesized CPWs, the on-chip dual-mode rat-race coupler is realized, and the chip photo is shown in Fig. 5.34. The size is $3.234 \times 1.394\ \text{mm}^2$. Figure 5.35(a) plots the *S*-parameters in both bands with the sum port (port 1) being excited. At the center frequency of the low band (2.75 GHz), the measured reflection coefficient ($|S_{11}|$) and isolation ($|S_{41}|$) are below −25 dB, while the transmission coefficients ($|S_{21}|$ and $|S_{31}|$) are −4.3 and −4.5 dB. The phase differences in the sum mode ($\angle S_{21} - \angle S_{31}$) and difference mode ($\angle S_{24} - \angle S_{34}$) are plotted in Fig. 5.35(b). They are −1.4° and −183.6°, respectively, at the center frequency.

Referring back to Fig. 5.35(a), the four-port network becomes a pair of uncoupled transmission lines (paths 1–2 and 3–4) in the high band. At 5.6 GHz, the measured reflection coefficient ($|S_{11}|$) is −34 dB while the transmission coefficient ($|S_{21}|$) is −2.4 dB. The remaining two ports, ports 3 and 4, are isolated from the transmission path 1–2. The $|S_{31}|$ and $|S_{41}|$ are at least −30 dB. Similar responses can be observed when the port 4 is excited and the path 3–4 is the transmission channel.

Last but not the least, after reading through the chapters, the readers may have noticed that the transmission line portion is a necessary part of the synthesized line if the design specifications involve different line impedances at different frequencies. In essence, the transmission line is an impedance transformer whose electrical length is *not* equal to 90°.

5.5.3 *Triplexer*

An on-chip triplexer without the need of an input matching section is then designed using the tri operational mode synthesized lines in Sec. 3.4 (Categories 3 and 4) [15]. As shown in Fig. 5.36, there are three circuit blocks, all serving as a frequency selective element and a matching unit at the same time. Each synthesized CPW (SCPW) behaves identical to a well-matched transmission line in one of the passbands, but resembles an open circuit in the other two. Benefiting from the open circuit condition with infinite input impedance, the inactive channels are completely hidden from the active one, making the synthesized CPWs able to be jointed together at the input terminal without accessory matching components. In this specific design, the operating bands are allocated at 1.8 (f_L), 2.6 (f_M), and 3.5 GHz (f_H).

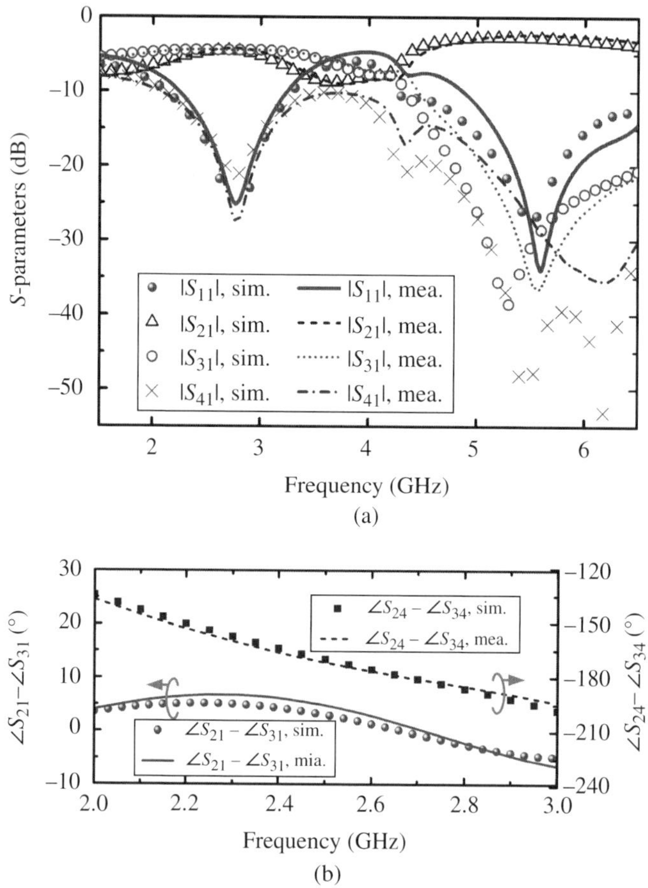

Figure 5.35 (a) *S*-parameters and (b) phase imbalances of the dual-mode rat-race coupler

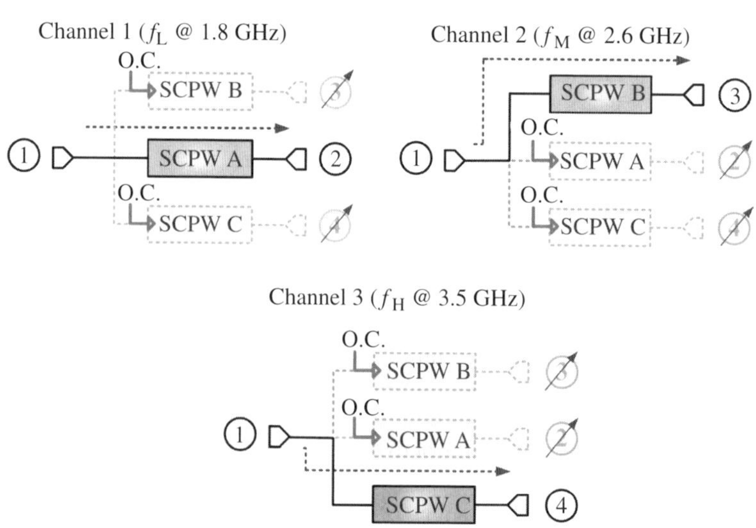

Figure 5.36 Schematic of the on-chip triplexer using synthesized CPWs (SCPWs)

By utilizing the IPD process, each synthesized CPW is realized by a quasi-lumped network composed of dual *LC* resonators and a simple L-section as the last stage. Referring to the ideal lumped model in Fig. 5.37(a), the resonators are in series connection, and each is associated with a transmission zero allocated at one of the operating frequencies (f_L, f_M, or f_H). In the practical implementation, the resonator with a higher resonance frequency, which corresponds to the one with smaller lumped values and circuit size, should be placed at the first stage of the network to mitigate the parasitic distributed effect of the inductors. The resonance frequency is

$$f_{TZj} = \frac{\omega_{TZj}}{2\pi} = \frac{1}{2\pi\sqrt{L_{ij}C_{ij}}} \quad (i = a,b,c,\ j = 1,2). \tag{5.49}$$

The L-section network (Z_d, Y_d) appended to the other side of the resonators is responsible for creating the transmission channel at the third frequency. A more realistic model, by taking into account the parasitic shunt capacitances of the inductors, is presented in Fig. 5.37(b). Since the lumped elements are all small compared to the guided wavelength, the off-resonance resonators can be further simplified to an extra L-section with (Z_{eq}, Y_{eq}) using node exchange. The approximate model is given in Fig. 5.37(c). The impedance Z_{eq} and admittance Y_{eq} are the direct sums of total series impedances and shunt admittances of the resonators as

$$Z_{eq} = Z_{eq1} + Z_{eq2} = \frac{j\omega L_{i1}}{1-\omega^2/\omega_{TZ1}^2} + \frac{j\omega L_{i2}}{1-\omega^2/\omega_{TZ2}^2} \tag{5.50}$$

$$Y_{eq} = j\omega\left(C_{i3} + C_{i4} + C_{i5} + C_{i6}\right). \tag{5.51}$$

In (5.50), the non-resonating *LC* tank is equivalent to either an inductor or a capacitor depending on the operating frequency. To ensure that the entire network is identical to a slow-wave synthesized line in the designated passband, the shunt admittance in the appended L-section must be capacitive, which is denoted as C_y. The series impedance Z_d, in the meantime, could be either inductive or capacitive depending on the total impedance Z_{eq} as well as the required characteristic impedance Z_c of the whole network in the passband. Again, by equating the

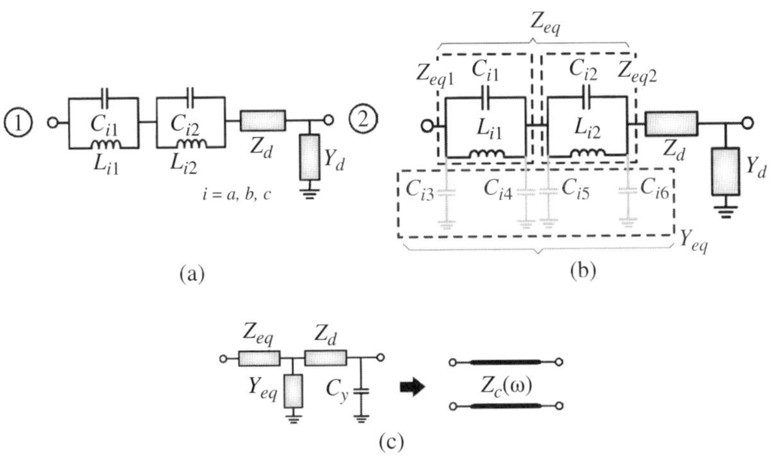

Figure 5.37 (a) Ideal, (b) realistic, and (c) simplified double L-section models of the synthesized CPW (SCPW). Source: Lai 2014 [15]. Reproduced with permission of IEEE

ABCD matrix of the synthesized CPW to that of a uniform transmission line, the characteristic impedance of the network is expressed as

$$Z_c(\omega) = \sqrt{\frac{B}{C}} = \sqrt{\frac{Z_d + Z_{eq} + Z_{eq}Y_{eq}Z_d}{1 + Y_{eq} + Y_{eq}Z_dY_d}}. \tag{5.52}$$

Equations (5.49) and (5.52) are the design equations of the synthesized CPW. As there are only three equations but six unknowns (L_{i1}, C_{i1}, L_{i2}, C_{i2}, Z_d, C_y), one has three degrees of freedom to be determined in accordance with layout considerations. In practice, the pair of inductors (L_{i1}, L_{i2}) could be selected first, which, in turn, determines the pair of capacitors (C_{i1}, C_{i2}) using (5.49). The shunt capacitor C_y in the appended L-section network yet provides another degree of freedom that can be arbitrarily selected. The final parameter Z_d is solved using (5.52) at the center frequency of the passband for a given characteristic impedance. Depending on the design requirements, this series impedance could be an inductor or a capacitor to make the overall network identical to a well-matched transmission line in the active channel. Note that since the network to realize the synthesized CPW is asymmetric, its electrical length cannot be determined using the *ABCD* matrix since A ≠ D. Fortunately, for the current design the electrical length is of no concern at all.

As a design example, Fig. 5.38 depicts the layout and *S*-parameters of the synthesized CPW A (SCPW A). At 1.8 GHz (f_L), the $|S_{11}|$ and $|S_{21}|$ are −19.2 and −1.3 dB, while they are −1.3 and

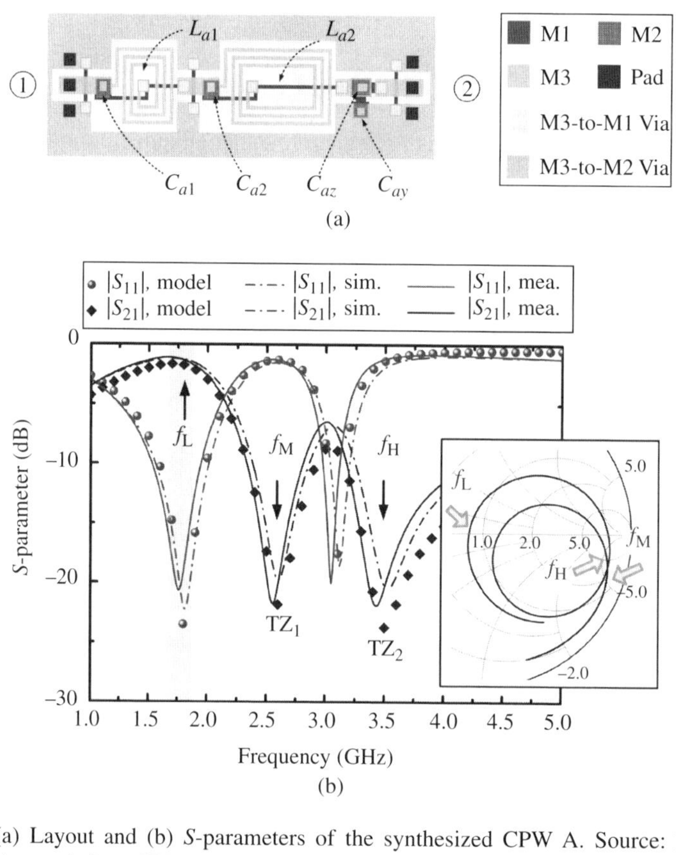

Figure 5.38 (a) Layout and (b) *S*-parameters of the synthesized CPW A. Source: Lai 2014 [15]. Reproduced with permission of IEEE

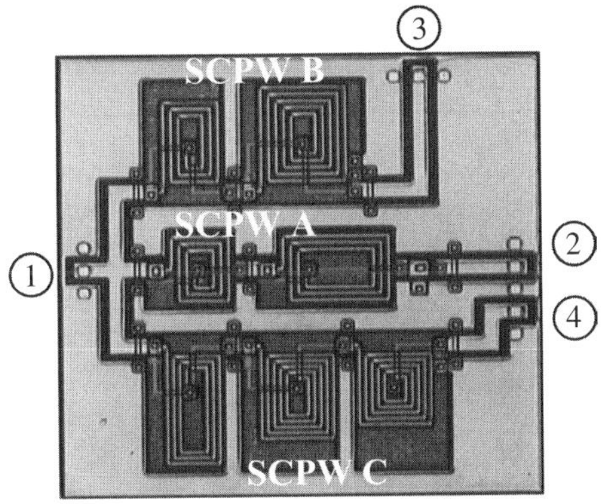

Figure 5.39 Chip photo of the on-chip triplexer

−20.2 dB at 2.6 GHz (f_M), and −1.2 and −20.4 dB at 3.5 GHz (f_H). The input reflection coefficients are $0.86\angle{-20.7^\circ}$ and $0.87\angle{-16.2^\circ}$ at f_M and f_H. Both values are close to the ideal open circuit condition ($1\angle 0^\circ$). The characteristic impedance is 48.8 Ω at the center frequency of the passband. From 1.73 to 1.9 GHz, the variation of the impedance is less than 20% (i.e., 50±10 Ω).

The on-chip triplexer is realized by a direct integration of the three synthesized CPWs without using additional matching sections. The chip photo is shown in Fig. 5.39. The circuit size is merely $1.4 \times 1.746\ \text{mm}^2$ ($0.015\,\lambda_g \times 0.0187\,\lambda_g$). Figure 5.40(a) depicts the reflection and transmission coefficients. It is well matched in the three bands. In the transmission channels, the power dissipations are $|S_{21}| = -1.9$ dB at f_L, $|S_{31}| = -2.2$ dB at f_M, and $|S_{41}| = -2.4$ dB at f_H. The input-to-inactive channel suppression is better than 25 dB at 1.8 GHz ($|S_{31}|$, $|S_{41}|$), 23 dB at 2.6 GHz ($|S_{21}|$, $|S_{41}|$), and 23 dB at 3.5 GHz ($|S_{21}|$, $|S_{31}|$). Figure 5.40(b) shows the output matching and isolations between the output ports. Over the three bands of concern, the isolation is at least 25 dB and could be up to 40–50 dB; the output reflection coefficients, meanwhile, are less than −13.5 dB.

The readers should be aware that by replacing the shunt admittance at the last stage with a series *LC* tank, a quad-mode synthesized line, that is, a line with a passband at one frequency but identical to an open circuit at all other three, is also feasible. The quad-mode line can be used to realize an on-chip quadruplexer, quad-mode phased array, and so on.

5.5.4 On-Chip Liquid Detector

Having a synthesized line with distinct electrical responses at two different operating frequencies is not the only way to fulfill dual-mode operation. In this subsection, instead of allocating the two operational modes at different frequencies, a new design by associating the two modes with the material medium by which the chip is surrounded is realized and investigated. It can be applied to determine the complex relative dielectric constant of a liquid or so.

The synthesis starts with the lumped circuit models in Fig. 5.41(a) and (b). Although at the first glance, it is identical in topology to the five-element ladder-type networks used throughout the book, the operational principle is quite different. First of all, referring to Fig. 5.41(a), the synthesized line should be equivalent to a uniform transmission line with (Z_a, θ_a) when the

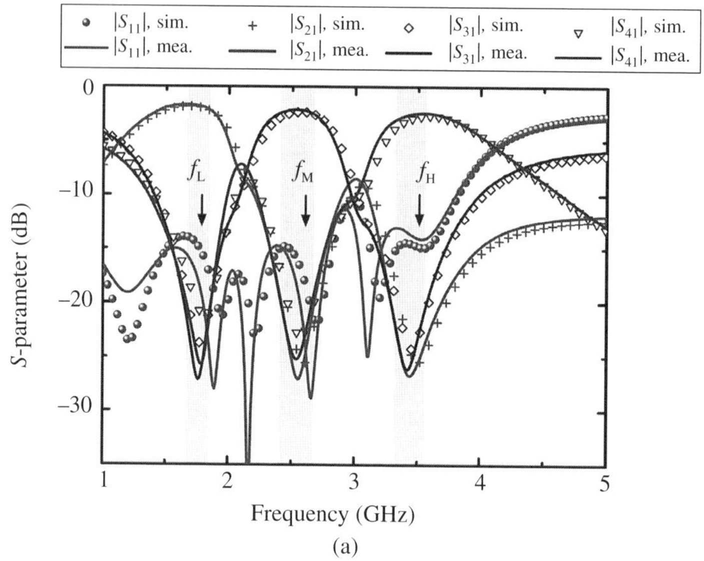

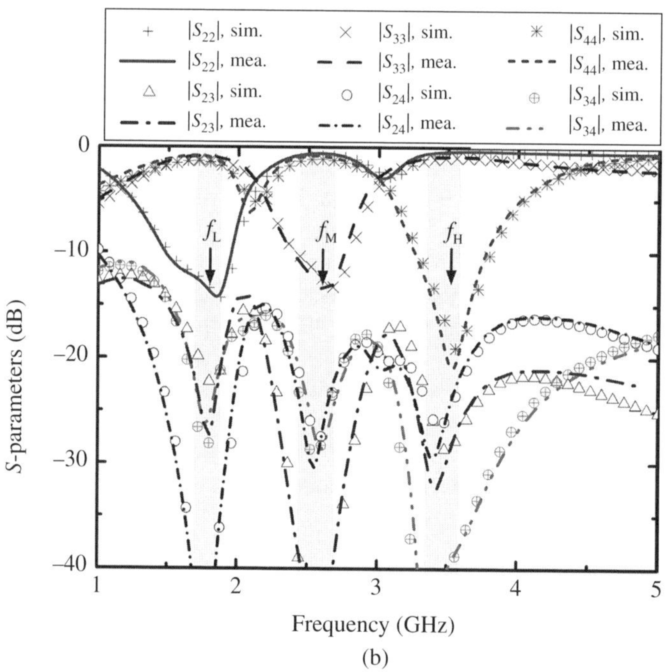

Figure 5.40 Performance of the on-chip triplexer: (a) reflection and transmission coefficients and (b) output matchings and port-to-port isolations. Source: Lai 2014 [15]. Reproduced with permission of IEEE

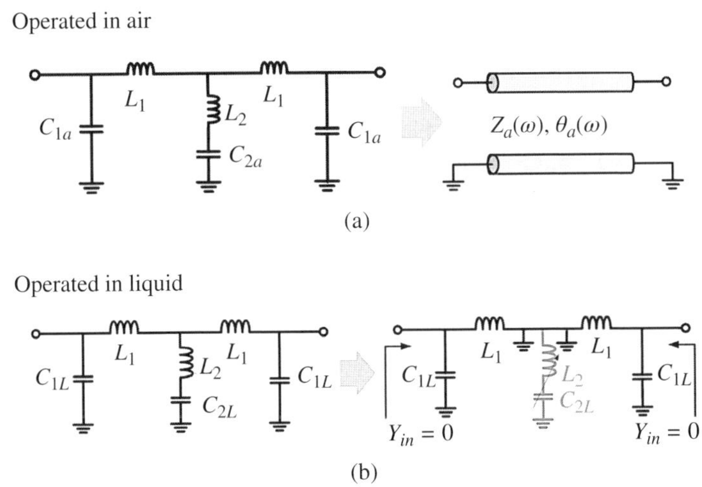

Figure 5.41 Lumped circuit models for realizing dual-mode synthesized CPW subject to the change of surroundings: (a) in air and (b) in liquid

network is surrounded by air at a given frequency f_c. By equating the *ABCD* matrix of the network to that of a uniform transmission line, one gets

$$Z_a(\omega) = \sqrt{\frac{B(\omega)}{C(\omega)}} = \sqrt{\frac{L_1\left(C_{2a}\left(L_1 + 2L_2\right)\omega^2 - 2\right)}{\left(1 - \omega^2 C_{1a} L_1\right)\left(-C_{2a} + C_{1a}\left(C_{2a}\left(L_1 + 2L_2\right)\omega^2 - 2\right)\right)}} \tag{5.53}$$

and

$$\theta_a(\omega) = \cos^{-1} A(\omega) = \cos^{-1}\left\{\frac{C_{2a}\left(L_1 + L_2\right)\omega^2 - \omega^2 C_{1a} L_1\left(-2 + C_{2a}\left(L_1 + 2L_2\right)\omega^2\right) - 1}{\omega^2 C_{2a} L_2 - 1}\right\}, \tag{5.54}$$

which are the characteristic impedance and electrical length of the synthesized line when operated in air.

Next, consider the synthesized line operated in liquid. The key factor behind this new idea is to control the values of capacitors in accordance with the surroundings. Referring to Fig. 5.41(b), by immersing the synthesized line into liquid, the capacitance values of the pair of capacitors are increased from (C_{1a} and C_{2a}) to (C_{1L} and C_{2L}); the inductance, meanwhile, is irrelevant to the change. The design goal is to make the synthesized line have infinite input impedance at the same operating frequency f_c when surrounded by liquid. Accordingly, the series *LC* tank should resonate and create a transmission zero at this frequency as

$$\omega_{TZ,L} = \omega_c = \frac{1}{\sqrt{L_2 C_{2L}}}. \tag{5.55}$$

Again, the series tank is short-circuited when resonating according to (5.55), which, in turn, forces the inductance L_1 and capacitance C_{1L} to form a parallel LC tank at each side of the network. By equating the resonance frequency of the parallel LC tank to that of the series one,

$$\frac{1}{\sqrt{L_1 C_{1L}}} = \frac{1}{\sqrt{L_2 C_{2L}}} = \omega_{TZ,L}, \tag{5.56}$$

the synthesized line is simply open-circuited at both ends.

The final piece to derive the design equation is to define the transformation ratio k as

$$k = \frac{C_{1L}}{C_1} = \frac{C_{2L}}{C_2}, \tag{5.57}$$

In short, k is the ratio of the value of a capacitor in liquid to that in air. It is a function of the layout as well as the substrate used, and can be evaluated by a full-wave simulator for a particular design. By substituting (5.57) into (5.55) and (5.56), we have

$$\frac{1}{\sqrt{L_1 C_{1a}}} = \sqrt{k}\omega_c = \omega_{TZ,a} \tag{5.58a}$$

and

$$\frac{1}{\sqrt{L_2 C_{2a}}} = \sqrt{k}\omega_c = \omega_{TZ,a}. \tag{5.58b}$$

$\omega_{TZ,a}$ is the angular frequency at which the transmission zero of the synthesized line should be allocated when surrounded by air. This transmission zero will be shifted down to the center frequency of the operating band (f_c) with infinite input impedance at both sides of the line when immersed in liquid.

Equations (5.53), (5.54), (5.58a), and (5.58b) form a set of simultaneous equations for the dual-mode synthesized transmission line. Since the number of equations and unknowns is the same, all lumped values can be uniquely solved for a targeted characteristic impedance, electrical length, and transmission zero frequency. The calculated values are then converted into physical footprints and integrated as a chip.

Following the design rule, a dual-mode synthesized CPW subject to the change of surroundings is developed and shown in Fig. 5.42(a). The operating frequency is 433 MHz, and the chip size is 3.7 × 1.97 mm^2 (0.00534 × 0.00284 λ_0^2). Here, the liquid is set as water, which has a relative dielectric constant of 81. A photograph of the test kit is shown in Fig. 5.42(b). The synthesized CPW can be used, for example, as an on-chip liquid detector for extracting the complex relative dielectric constant of an unknown liquid. Referring to (5.58), the resonance frequency of the LC tank is inversely proportional to the square root of the relative dielectric constant. With a small drop of an unknown liquid on the chip, the transmission zero

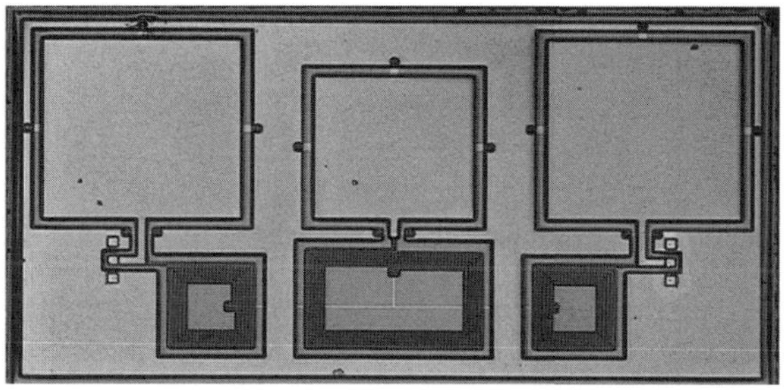

(a)

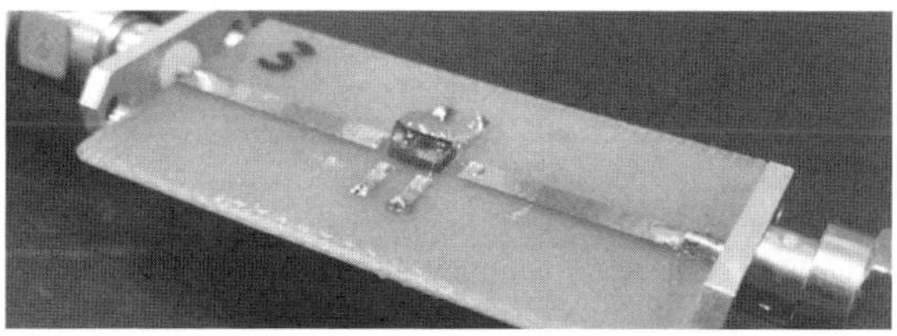

(b)

Figure 5.42 (a) Chip photo of the dual-mode synthesized CPW subject to the change of surroundings and (b) test kit with the chip mounted on a PCB; a drop of water covers the top surface of the chip

frequency shifts accordingly, and so does the insertion loss level at that frequency. The complex dielectric constant, transmission zero frequency, and insertion loss can be mathematically modeled by the following equations [25],

$$\Delta f = K_1 \Delta\varepsilon' + K_2 \Delta\varepsilon'' \tag{5.59}$$

$$\Delta|S_{21}| = K_3 \Delta\varepsilon' + K_4 \Delta\varepsilon''. \tag{5.60}$$

Δf is the amount of shift in the transmission zero frequency; $\Delta|S_{21}|$ is the variation of the insertion loss level; K_1 to K_4 are the sensitivity coefficients.

As a proof of concept, alcoholic solutions with different volume percentages of ethanol are applied to verify the effectiveness of the on-chip liquid detector. The transmission zero frequency and insertion loss when operated in air are set as the reference values. After a number of experiments and data analysis, the predicted complex relative dielectric constant is summarized in Fig. 5.43. The results from [26] are plotted for comparison. The water volume fraction of the solution varies from 0 to 100%. The agreement between the results is fairly good. The real part deviates from the predicted value only when the relative dielectric constant is very high such that the transmission zero becomes less sensitive to the variation of the material medium. The imagery part, on the other hand, is higher than expected. It is likely a result of the unaccounted parasitic loss due to the silicon substrate not calibrated out during measurement. Despite the small discrepancy, the concept of fulfilling a dual-mode synthesized CPW subject to change of surroundings is clearly demonstrated. Readers are encouraged to find more innovative applications by using this sort of synthesized transmission lines!

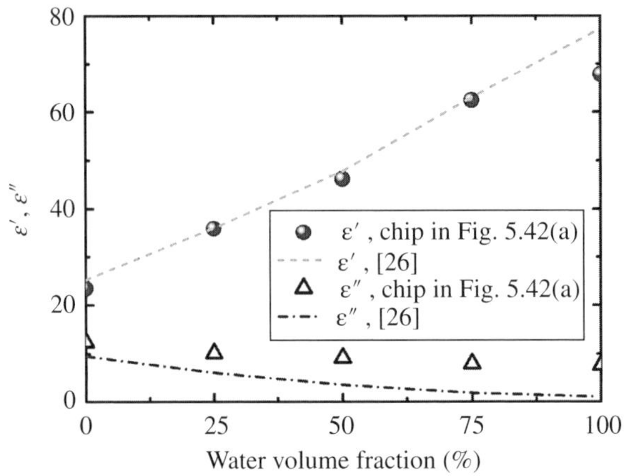

Figure 5.43 Estimation of the real part (ε′) and imaginary part (ε″) of the relative dielectric constant of an alcoholic solution with different water volume fractions

References

[1] I. Haroun, C. Plett, Y.-C. Hsu, and D.-C. Chang, "Compact 60-GHz IPD-based branch-line coupler for system-on-package V-band radios," *IEEE Trans. Comp., Pack. Manuf. Techn.*, vol. 2, no. 7, pp. 1070–1074, July 2012.

[2] Y.-T. Chen, C.-L. Chang, and C.-H. Tseng, "A compact X-band CPW branch-line coupler using glass integrated passive device (GIPD) technology," *in Proc. IEEE 2013 Asia-Pacific Microw. Conf. (APMC)*, Seoul, Korea, Nov. 5–8, 2013, pp. 276–278.

[3] Y.-C. Hsu, I. Haroun, D.-C. Chang, R. Amaya, and K. Hettak, "A low-cost small-size 60 GHz quadrature coupler for IPD-based SiP radio systems," *in Proc. IEEE 2014 Asia-Pacific Microw. Conf. (APMC)*, Sendai, Japan, Nov. 4–7, 2014, pp. 831–833.

[4] D. Titz, R. Pilard, C. Laporte, S. Jan, F. Gianesello, D. Gloria, G. Jacquemod, and C. Luxey, "New wideband miniature branchline coupler on IPD technology for beamforming applications," *IEEE Trans. Comp., Pack. Manuf. Techn.*, vol. 4, no. 5, pp. 911–921, May 2014.

[5] I. Haroun, Y.-C. Hsu, and D.-C. Chang, "60-GHz rat-race coupler using LG-CPW transmission lines in IPD technology," *in Proc. IEEE 2010 Asia-Pacific Microw. Photonics Conf.*, Singapore, Oct. 18–21, 2011, pp. 284–287.

[6] Y.-C. Tseng and T.-G. Ma, "On-chip miniaturized 3-dB directional coupler using coupled synthesized CPWs on integrated passive device (IPD) process," *in Proc. 43th Eur. Microw. Conf.*, 2014, Rome, Italy, Oct. 6–9, 2014, pp. 81–84.

[7] C.-H. Huang, T.-C. Wei, T.-S. Horng, S.-M. We, C.-C. Wang, C.-T. Chiu, and C.-P. Hung, "High-performance Marchand-type balun design and fabrication using an integrated passive device (IPD) technology," *in Proc. 2008 Elect. Materials Pack. Conf. (EMAP)*, Taipei, Taiwan, Oct. 22–24, 2008, pp. 137–140.

[8] Y.-S. Lin and J.-H. Lee, "Miniature Butler matrix design using glass-based thin-film integrated passive device technology for 2.5-GHz applications," *IEEE Trans. Microw. Theory Techn.*, vol. 61, no. 7, pp. 2594–2602, July 2013.

[9] H.-C. Lu, C.-S. Yeh, S.-A. Wei, and Y.-T. Chou, "60 GHz CPW dual-mode rectangular ring bandpass filter using integrated passive devices process," *in Proc. IEEE 2010 Asia-Pacific Microw. Conf. (APMC)*, Yokohama, Japan, Dec. 7–10, 2010, pp. 1883–1886.

[10] C.-Y. Hsiao, S. S.H. Hsu, and D.-C. Chang, "A compact V-band bandpass filter in IPD technology," *IEEE Microw. Wireless Comp. Lett.*, vol. 21, no. 10, pp. 531–533, Oct. 2011.

[11] Y.-C. Tseng and T.-G. Ma, "On-chip GIPD bandpass filter using synthesized stepped impedance resonators," *IEEE Microw. Wireless Comp. Lett.*, vol. 24, no. 3, pp. 140–142, March 2014.

[12] C.-H. Chen, C.-H. Huang, T.-S. Horng, S.-M. Wu, C.-T. Chiu, C.-P. Hung, J.-Y. Li, and C.-C. Chen, "Very compact transformer-coupled balun-integrated bandpass filter using integrated passive device technology on glass substrate," in *IEEE MTT-S Int. Microwave Symp. Dig.*, Anaheim, CA, pp. 1372–1375, May 23–28, 2010.

[13] C.-Y. Hsiao, Y.-C. Huang, and T.-L. Wu, "An ultra-compact common-mode bandstop filter with modified-T circuits in integrated passive device (IPD) process," *IEEE Trans. Microw. Theory Techn.*, vol. 63, no. 11, pp. 3624–2631, Nov. 2015.
[14] C.-H. Lai, G.-T. Zhou, and T.-G. Ma, "On-chip miniaturized diplexer using jointed dual-mode right-/left-handed synthesized coplanar waveguides on GIPD process," *IEEE Microw. Wireless Comp. Lett.*, vol. 24, no. 4, pp. 245–247, April 2014.
[15] C.-H. Lai, W.-S. Chung and T.-G. Ma, "On-chip miniaturized triplexer using lumped networks with dual resonators on integrated passive device process," *IEEE Trans. Microw. Theory Techn.*, vol. 62, no. 12, pp. 5992–5999, Dec. 2014.
[16] S. Wang and W.-J. Lin, "A 10/24-GHz CMOS/IPD monopulse receiver for angle-discrimination radars," *IEEE Trans. Circuit Systems-I*, vol. 61, no. 10, pp. 2999–3006, Oct. 2014.
[17] H.-J. Wei, C.-C. Meng, S.-W. Yu, and C.-H. Chang, "A Chebyshev-response step-impedance phase-inverter rat-race coupler directly on lossy silicon substrate and its Gilbert mixer application," *IEEE Trans. Microw. Theory Techn.*, vol. 59, no. 4, pp. 882–893, Apr. 2011.
[18] T.-T. Mo, Q. Xue, and C.-H. Chan, "A broadband compact microstrip rat-race hybrid using a novel CPW inverter," *IEEE Trans. Microw. Theory Techn.*, vol. 55, no. 1, pp. 161–167, Jan. 2007.
[19] C.-L. Hsu, C.-W. Chang, and J.-T. Kuo, "Miniaturized dual-band hybrid couplers with arbitrary power division ratios," *IEEE Trans. Microw. Theory Techn.*, vol. 57, no. 1, pp. 149–156, Jan. 2009.
[20] C.-H. Tseng, "Compact LTCC rat-race couplers using multilayered phase-delay and phase-advance T-equivalent sections," *IEEE Trans. Adv. Packag.*, vol. 33, no. 2, pp. 543–551, May. 2010.
[21] M. Ercoli, D. Dragomirescu, and R. Plana, "An extremely miniaturized ultra wide band 10–67 GHz power splitter in 65 nm CMOS technology," *in IEEE MTT-S Int. Microw. Symp. Dig.*, 2012.
[22] J. S. Hong and M. L. Lancaster, *Microstrip Filters for RF/Microwave Applications*, 1st edn, John Wiley & Sons, Ltd, 2001.
[23] R. K. Mongia, I. J. Bahl, P. Bhartia, and J. S. Hong, *RF and Microwave Coupled-line Circuits*, 2nd edn, Artech House, 2007.
[24] K.-C. Lin, C.-H. Wu, C.-H. Lai, and T.-G. Ma, "Novel dual-band decoupling network for two-element closely spaced array using synthesized microstrip lines," *IEEE Trans. Microw. Theory Techn.*, vol. 60, no. 11, pp. 5118–5128, Nov. 2012.
[25] T. Chretiennot, D. Dubuc, and K. Grenier, "A microwave and microfluidic planar resonator for efficient and accurate complex permittivity characterization of aqueous solutions," *IEEE Trans. Microw. Theory Techn.*, vol. 61, no. 2, pp. 972–978, 2013.
[26] A. Megriche, A. Belhadj, A. Mgaidi, "Microwave dielectric properties of binary solvent water-alcohol, alcohol-alcohol mixtures at temperatures between −35°C and +35°C and dielectric relaxation studies," *Mediterranean Journal of Chemistry.*, vol. 1, no. 4, pp. 200–209, 2012.

6

Periodic Synthesized Transmission Lines with Two-Dimensional Routing

T. G. Ma

6.1 Introduction

As introduced in Chapter 1, synthesized transmission lines can be categorized into two groups: periodic structures [1–6] and non-periodic structures [7–24]. Non-periodic synthesized lines, intensively studied in previous chapters, are based on lumped/quasi-lumped networks capable of mimicking a uniform transmission line with a large amount of phase delay (say, 90°). The network could be either right-handed or left-handed [11] and the equivalence is generally on a narrowband basis. The $\pi-$and T-networks are among the most popular configurations to achieve the goal [11, 12, 15, 16]. Non-periodic synthesized lines with harmonic suppression and dual- or even tri-mode operations have been investigated and demonstrated [8, 9, 24]. The bridged T-coil or the modified-T/π network with coupled inductors provides extraordinarily compact footprints among the designs [13, 18, 22].

In the meantime, synthesized lines realized by periodic unit cells have also been studied to realize a transmission line with an arbitrary electrical length and a variety of characteristic impedances. In these designs, each unit cell, while kept at a constant impedance level, is just associated with a small electrical length. Host lines loaded by periodic capacitive stubs are developed for compact hybrid couplers in [1]. Periodic interdigital stubs served as the mutual capacitance of a forward-wave directional coupler are introduced in [2]. The complementary-conducting-strip transmission lines (CCS TLs) offer two-dimensional routing capability while preserving acceptable power dissipation even on the lossy CMOS process [3–6]. In general, periodic synthesized transmission lines provide systematic design procedure, moderate-to-good size reduction ratio, and quite a flexible layout arrangement.

Synthesized Transmission Lines: Design, Circuit Implementation, and Phased Array Applications, First Edition.
Tzyh-Ghuang Ma, Chao-Wei Wang, Chi-Hui Lai and Ying-Cheng Tseng.

In this chapter, a new sort of one-dimensional periodic synthesized transmission lines with two-dimensional routing capability is introduced using the printed circuit board (PCB) technology. The fundamental building block is a square unit cell composed of crossed conducting strips and metallic pads interleaved between strips. An additional patch centered at the unit cell could be incorporated to adjust the impedance level of the line. Two out of the four conducting strips are selected as the input/output ports, making the design fulfill two-dimensional layout routing.

Three unit cells, each with a given characteristic impedance, are developed and cascaded as 90° lines. Although the layout pattern, at the first glance, is similar to that of the PBG structure in Sec. 1.6 [25], the design methodologies are quite different. The periodic cells in this chapter are etched on the signal traces to function as transmission lines with low-pass nature, while those in the PBG are designed to cover the entire ground plane and serve mainly as a bandgap structure.

The quarter-wavelength periodic synthesized lines are then applied to realize a number of dividers and couplers, including a Wilkinson power divider, branch-line coupler, rat-race coupler, and broadside-coupled directional coupler. All designs show very compact sizes but are with good performance. A common-mode rejection filter is also developed using multilayered PCB process. Finally, on-chip implementation on the integrated passive device (IPD) substrate is introduced and analyzed.

6.2 Design of the Unit Cells

6.2.1 Formulation

Figure 6.1(a) and (b) shows the circuit layout and corresponding lumped model of the unit cell for realizing the periodic synthesized transmission line with two-dimensional routing capability. Here a 50-Ω line is given as an example. The unit cell is developed on a 20-mil Roger RO4003C substrate. The center operating frequency is again 900 MHz for easy comparison.

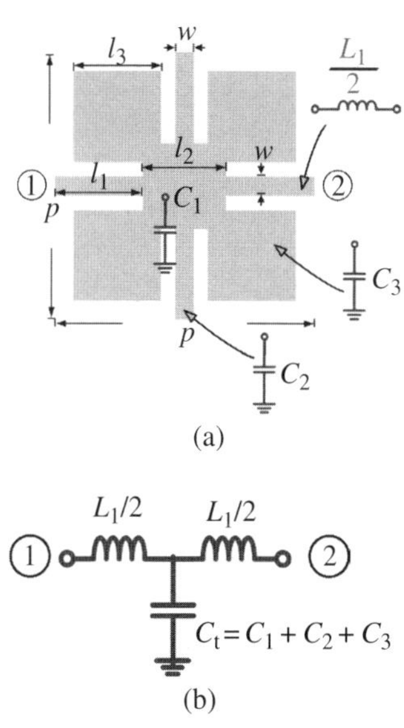

Figure 6.1 (a) Circuit layout and (b) equivalent lumped circuit model of the unit cell for the periodic synthesized transmission line ($l_1 = l_3 = 0.95$ mm, $l_2 = 0.9$ mm, $w = 0.2$ mm)

The unit cell is in square shape. The side of the square, or the periodicity, is p. A pair of crossed conducting strips is placed at the center of the cell with a square patch overlapped with it. The crossed strips divide the space into four quadrants, each filled with an additional capacitive pad; the pad is electrically connected the center patch using one of its four corners. Two out of the four conducting strips serve as the connecting lines to the input/output ports, and hence provide the series inductance (L_1). The other two strips, together with the pads at the center and four quadrants, fulfill the capacitive part (C_1, C_2, and C_3) of the unit cell. It should be noted that, although in the figure the input/output ports are opposite to each other as in a conventional transmission line, an alternative arrangement with the I/O ports placed at a right angle to one another is also possible. It therefore facilitates two-dimensional routing without additional effort.

Since the capacitive pads/stubs are all jointed together at the center of the cell, the equivalent lumped circuit can be as simple as a T-model as shown in Fig. 6.1(b). Here, $L_1/2$ is the series inductance sourced by one of the connecting arms, and C_t is the direct sum of all capacitances shunted to ground.

For a low impedance line, the patch at the center could be enlarged to increase the shunt capacitance and hence decrease the series inductance accordingly. On the other side, for a high impedance line, the center patch could be removed to reduce the total shunt capacitance. It, in turn, increases the length of the connecting arms and hence their corresponding series inductance. The capacitive pad in each quadrant could be replaced by an open stub to further reduce the shunt capacitance if required.

The propagation characteristic of the unit cell is analyzed by applying the standard Bloch wave analysis as introduced in Sec. 1.3.1. In Fig. 6.2, with an infinite number of unit cells connected in cascade, the voltage/current relationship between the nodes n and $n+1$ can be expressed by the $ABCD$ matrix of the unit cell as

$$\begin{bmatrix} V_n \\ I_n \end{bmatrix} = \begin{bmatrix} 1+Z_{eq}Y_{eq} & Z_{eq}\left(1+Z_{eq}Y_{eq}\right)+Z_{eq} \\ Y_{eq} & 1+Z_{eq}Y_{eq} \end{bmatrix} \begin{bmatrix} V_{n+1} \\ I_{n+1} \end{bmatrix} \tag{6.1}$$

where

$$Z_{eq} = j\omega L_1/2 \tag{6.2}$$

$$\mathrm{Y}_{eq} = j\omega C_t = C_1 + C_2 + C_3. \tag{6.3}$$

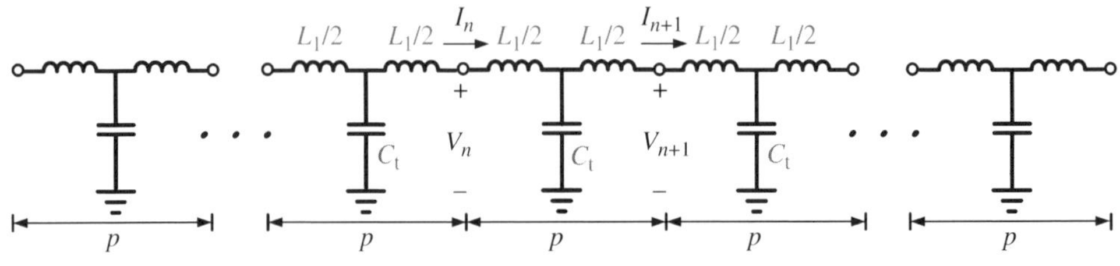

Figure 6.2 An infinite number of unit cells connected in cascade

From (1.37), the Bloch impedance is

$$Z_B = \frac{\pm B}{\sqrt{A^2-1}} = \sqrt{\frac{Z_{eq}\left(2+Z_{eq}Y_{eq}\right)}{Y_{eq}}}. \tag{6.4}$$

Since the unit cell is very short compared to the guided wavelength, the lumped inductance and capacitance in the cell are very small, and the cutoff angular frequency, which is represented by

$$\omega_c = \frac{2}{\sqrt{L_1 C_t}}, \tag{6.5}$$

is far away from the frequency band of interest. It suggests

$$Z_{eq}Y_{eq} \ll 2. \tag{6.6}$$

Substitution of (6.6) into (6.4) gives

$$Z_B = \sqrt{\frac{Z_{eq}\left(2+Z_{eq}Y_{eq}\right)}{Y_{eq}}} \approx \sqrt{\frac{2Z_{eq}}{Y_{eq}}} = \sqrt{\frac{L_1}{C_t}} = Z_c. \tag{6.7}$$

The simplified formula (6.7) is exactly in the same form as the one by directly equating the *ABCD* matrix of the unit cell to that of a uniform transmission line, a technique intensively used in previous chapters to deal with non-periodic synthesized transmission lines. Nevertheless, readers should be aware that the analysis bases are very different.

Meanwhile, from (1.35) and with reciprocal, symmetric and lossless properties, the phase constant, under the assumption of a very small electrical length, is

$$\beta_g \cong \sqrt{-2Z_{eq}Y_{eq}} \,/\, p = \omega\sqrt{L_1 C_t} \,/\, p. \tag{6.8}$$

Or alternatively, the electrical length of the unit cell is

$$\theta_p = \beta_g p = \omega\sqrt{L_1 C_t}. \tag{6.9}$$

Equations (6.7) and (6.9) are the design formulae for synthesizing the unit cell. The periodicity, or the size of the square p, should be chosen first. For a given impedance level and electrical length (which is only a few degrees), the required total inductance (L_1) and capacitance (C_t) can be solved from the simultaneous equations. While the width of the line inductor (w) is kept close to the minimum resolution of the fabrication process, the length (l_1) is selected to fix the pair of series inductance ($L_1/2$). This, in turn, determines the size of the center patch ($l_2 = p - 2l_1$) and its associated capacitance (C_1). The capacitance of the remaining two arms not connected to adjacent cells, C_2, are fixed at the same time. Finally, the area of the pad (l_3) in each quadrant is adjusted for a proper value of C_3 to fulfill the given total capacitance C_t, and hence complete the design.

6.2.2 *Quarter-Wavelength Lines*

By applying the aforementioned synthesis procedure, a 50-Ω, 90° synthesized transmission line using the periodic unit cell in Fig. 6.1 is developed and shown in Fig. 6.3(a). Figure 6.3(b) is a photograph of the fabricated sample. Each cell is designed with $Z_B = 50\ \Omega$ and $\theta_p = 9°$, implying that one needs 10 cells to fulfill the quarter-wavelength line. The periodicity is 2.8 mm. From (6.7) and (6.9), L_l and C_t are solved as 1.39 nH and 0.56 pF. The lumped values are converted to footprints and integrated as a whole. In the layout, the 90° line is meandered by connecting adjacent cells at right angles to reduce the physical length and hence utilize the circuit area more efficiently. Since the bend inevitably decreases the series inductance due to a slightly shorter propagation path, a pair of conventional microstrip lines can be appended to the outermost periodic unit cells for phase adjustment. Figure 6.4(a) shows the model-calculated, full-wave simulated, and measured *S*-parameters of the line. The equivalent periodic circuit model in Fig. 6.2 is used to calculate the model data. In the measurement, the TRL (thru-reflect-line) calibration is applied to calibrate out the SMA adaptor and extra microstrip lines for connecting purpose. The synthesized line is well matched to 50 Ω with the reflection coefficient lower than −30 dB over the band. Figure 6.4(b) shows the characteristic impedance and phase delay of the line. The characteristic impedance is extracted using (1.45). The measured data is 50.7 Ω and 84° at the center frequency of 900 MHz. The electrical length is slightly lower than the expected value due to parasitic coupling between cells in close proximity. In a practical design, an extra cell or a section of common microstrip line could be added to compensate for the difference.

Using a similar procedure, two more periodic synthesized lines with two-dimensional routing capability are developed. The Bloch impedance of the unit cells are 35 and 70 Ω, respectively. Figure 6.5(a) and (b) depicts the layouts of the unit cells. For the 35-Ω cell, the center patch is boosted as a large MIM capacitor to meet the need of a high capacitance in a low-impedance line. Meanwhile, the connecting inductive line is reduced in inductance by widening the line width (w) since the periodicity has been fixed as a constant value. The small

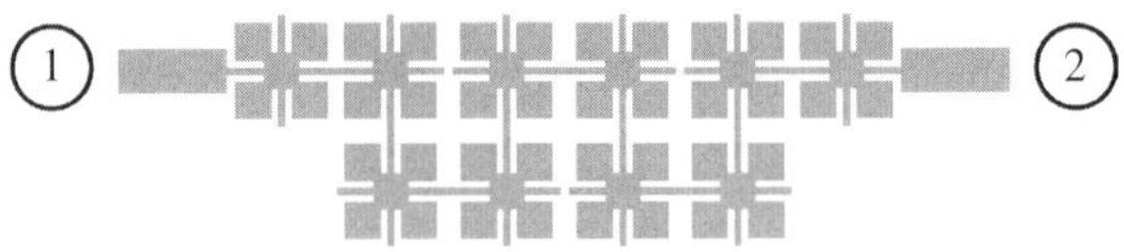

(a)

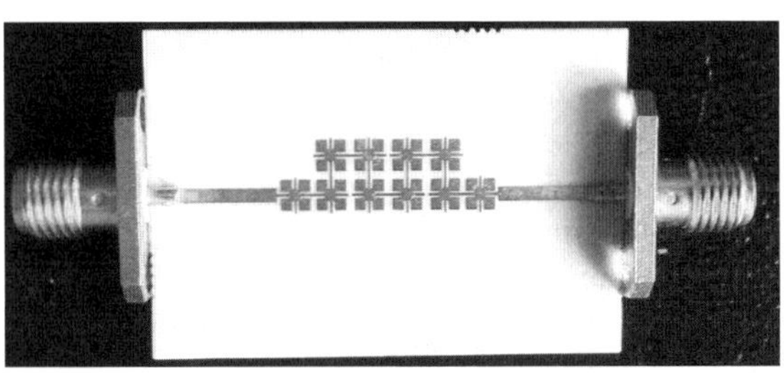

(b)

Figure 6.3 (a) Circuit layout and (b) photograph of 50-Ω, 90° synthesized line using the unit cells in Figure 6.1

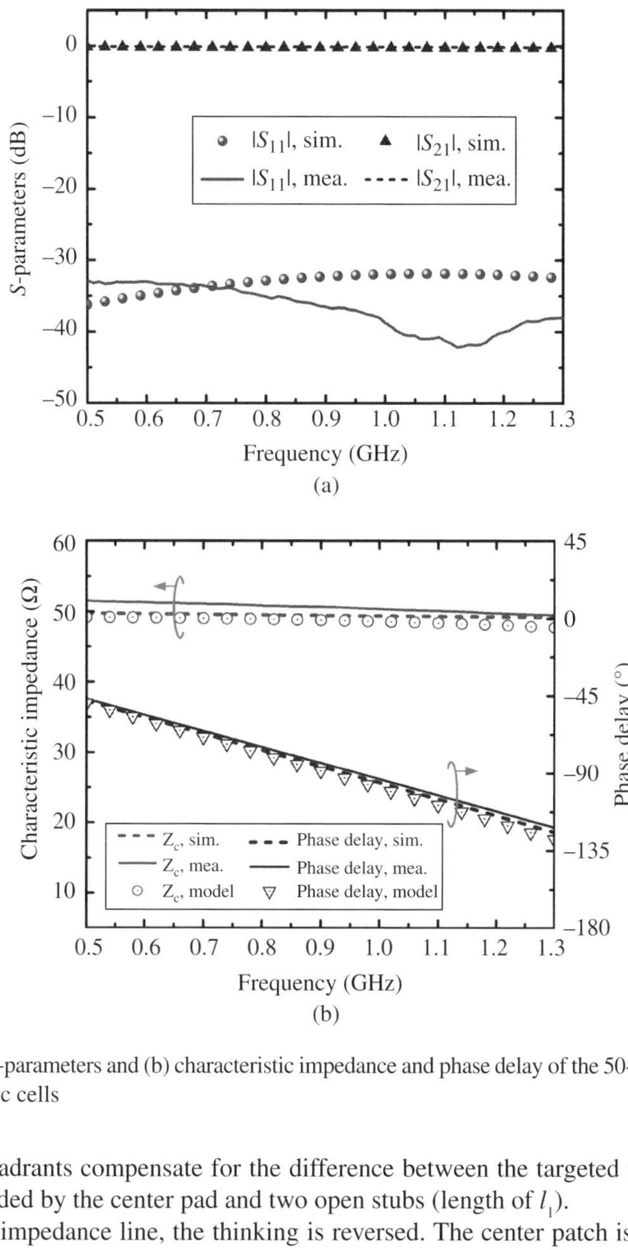

Figure 6.4 (a) *S*-parameters and (b) characteristic impedance and phase delay of the 50-Ω, 90° synthesized line using periodic cells

pads in four quadrants compensate for the difference between the targeted total capacitance and those provided by the center pad and two open stubs (length of l_1).

For the high-impedance line, the thinking is reversed. The center patch is shrunk to minimize the capacitance such that it is completely overlapped with the cross section of strips. The pads in four quadrants are also replaced by open stubs and the final layout reveals an asterisk shape. The line width, again, is chosen close to the minimum line resolution.

These unit cells are also connected in cascade as quarter-wavelength lines. The layouts of the 35-Ω, 90° and 70-Ω, 90° synthesized transmission lines are illustrated in Fig. 6.6(a) and (b). The numbers of cells are 16 and 24, respectively. The routing could be arbitrary but the current one makes the lines best fit to the coupler designs in Sec. 6.3. The occupied sizes of all lines,

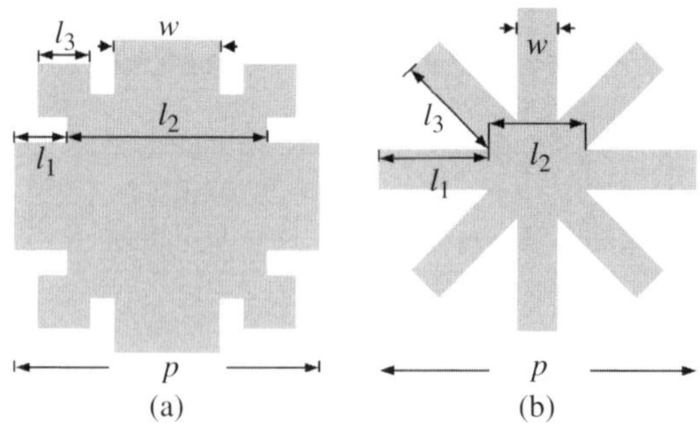

Figure 6.5 Circuit layouts of the (a) 35-Ω and (b) 70-Ω unit cells ($l_1 = l_3 = 0.45$ mm, $l_2 = 1.7$ mm, $w = 0.9$ mm for the 35-Ω cell, and $l_1 = l_3 = 0.55$ mm, $l_2 = 0.5$ mm, $w = 0.2$ mm for the 70-Ω one)

including the 50-Ω case, are around several hundredths of λ_g^2, and the slow wave factor is 2–3. The attenuation constant is around 0.3–0.5 dB/λ_g. When compared to that of a non-periodic synthesized transmission line (such as the CPW cases in Chapter 2), the slow wave factor of a synthesized line using periodic cells is by no means impressive. It is the two-dimensional routing capability makes this sort of synthesized transmission lines also features a compact size. It is easier to be filled into an arbitrarily given space than its non-periodic counterpart, as well. Moreover, since the unit cell is very small compared to the wavelength, the cutoff frequency is far away from the operating band, making the periodic synthesized line nondispersive in impedance and with linear phase response. If the linearity of a system is the first priority to be considered, one should select a proper periodic synthesized line for circuit miniaturization without causing detrimental harmonics.

6.3 Power Divider and Couplers

The quarter-wavelength synthesized lines in the previous section can be directly integrated together as power dividers and couplers without much tuning effort. First of all, the layout of a Wilkinson power divider is shown in Fig. 6.7. The two conventional 70-Ω 90° microstrip lines in the divider are replaced by synthesized ones introduced in Fig. 6.5(b). Different from the layout in Fig. 6.6(b), the unit cells are re-arranged such that both lines are in rectangular shape to show up the routing flexibility of this sort of design. When compared with its conventional counterpart, the size reduction ratio of the miniaturized Wilkinson power divider is up to 83%. Figure 6.8 shows the *S*-parameters of the divider. At 900 MHz, the $|S_{11}|$ is −27 dB, and the $|S_{21}|$ and $|S_{31}|$ are −3.2 and −3.15 dB, respectively. The amplitude and phase imbalances are negligible over a bandwidth of 70%.

The second example is a branch-line coupler, comprising pairs of 35-Ω and 50-Ω quarter-wavelength synthesized lines. The lines in Fig. 6.3 and 6.6(a) are directly jointed together to fulfill the goal. The layout is depicted in Fig. 6.9, in which the unit cells are tightly placed to get an overall compact footprint. It is only 20% the size of a typical microstrip design using uniform lines. Figure 6.10 shows the *S*-parameters of the branch-line coupler; the phase difference between the output ports is illustrated at the same time. At the center frequency, the reflection coefficient ($|S_{11}|$) is lower than −45 dB, while the isolation ($|S_{41}|$) is better than 35 dB.

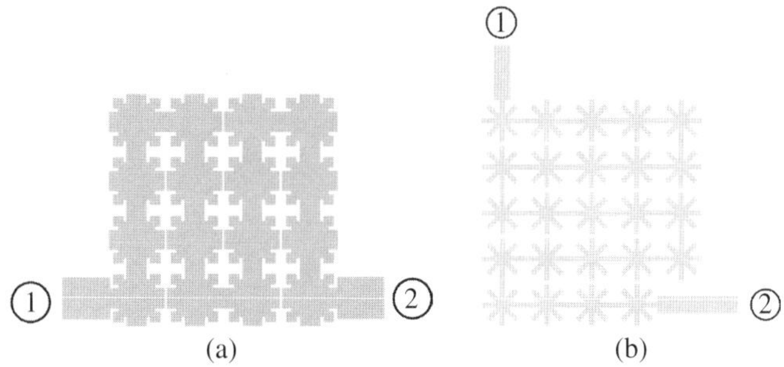

Figure 6.6 Circuit layouts of the (a) 35-Ω, 90° and (b) 70-Ω, 90° synthesized transmission lines using periodic cells in Fig. 6.5

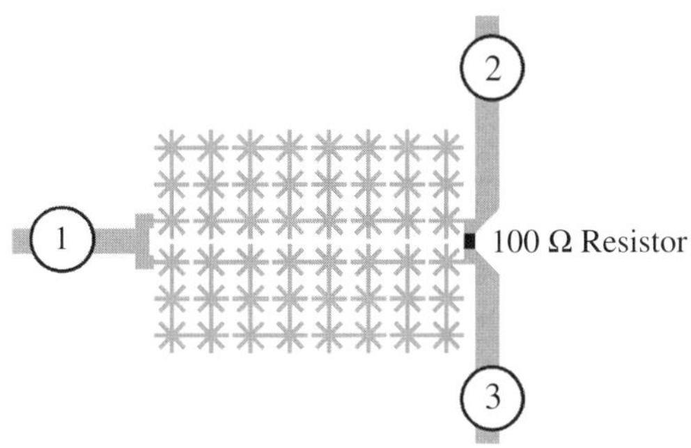

Figure 6.7 Layout of the Wilkinson power divider using periodic synthesized lines

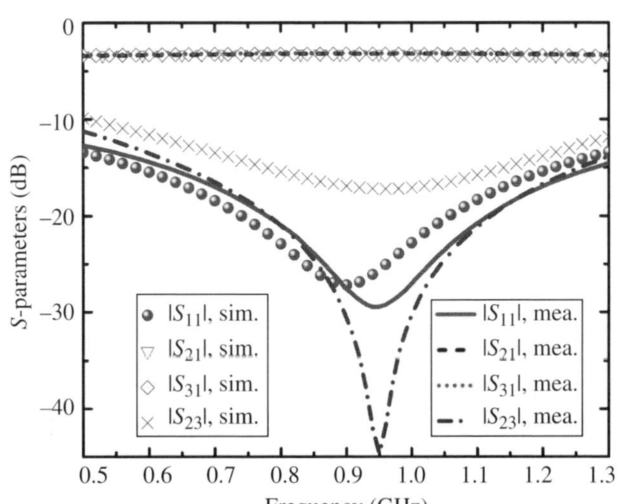

Figure 6.8 *S*-parameters of the Wilkinson power divider

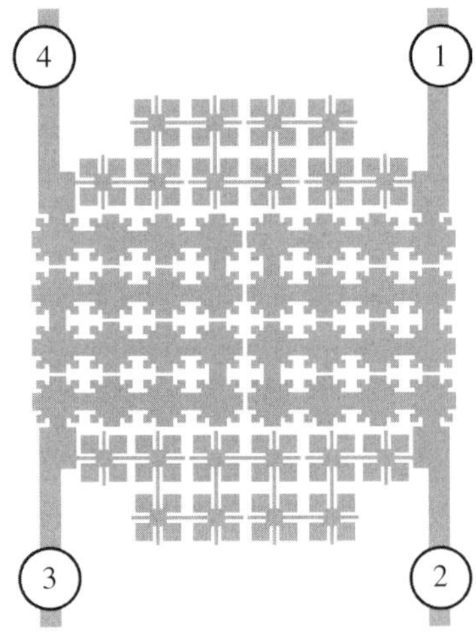

Figure 6.9 Layout of the branch-line coupler using periodic synthesized lines

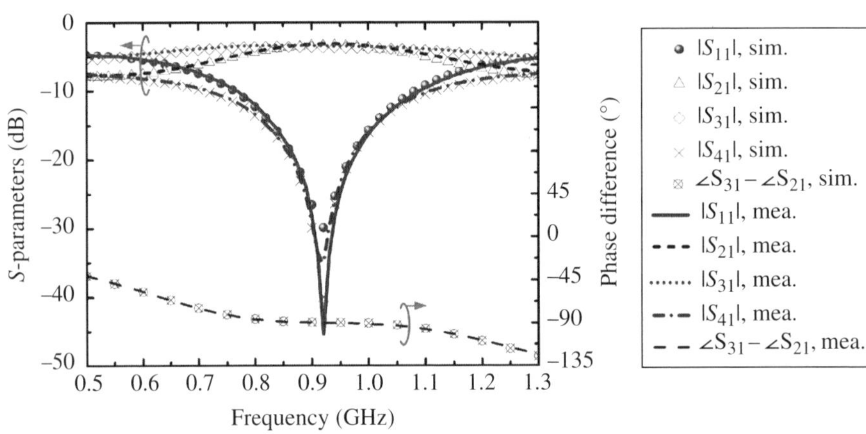

Figure 6.10 *S*-parameters and phase difference at the outputs of the branch-line coupler

The transmission coefficients ($|S_{21}|$, $|S_{31}|$) are −3.1 and −3.4 dB, respectively. The outputs are almost in phase quadrature, and the deviation from the ideal value is only 0.04°. The fractional bandwidth is about 20%, and the responses are symmetric with respect to the center frequency like a conventional design.

Figure 6.11 is the layout of the rat-race coupler. Three quarter-wavelength and one three-quarter wavelength synthesized lines are combined together to complete the design. As one may infer from the figure, there are many ways to route the 70-Ω unit cells to achieve a quarter-wavelength line. In a real integration, the periodic unit cells make the coupler easy to be filled out in an empty space with an arbitrary outline. When compared with a conventional 1.5-λg 180° ring coupler, the size reduction ratio is as high as 94%!

Figure 6.12(a) and (b) reveals the simulated and measured *S*-parameters and output phase differences when the sum and difference ports are excited, respectively. The reflection

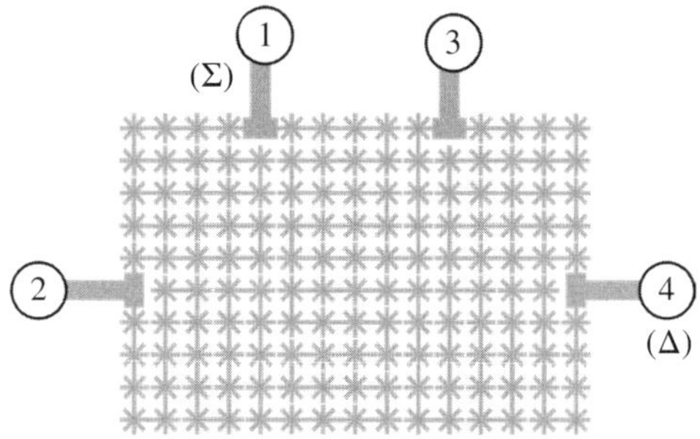

Figure 6.11 Layout of the rat-race coupler using periodic synthesized lines

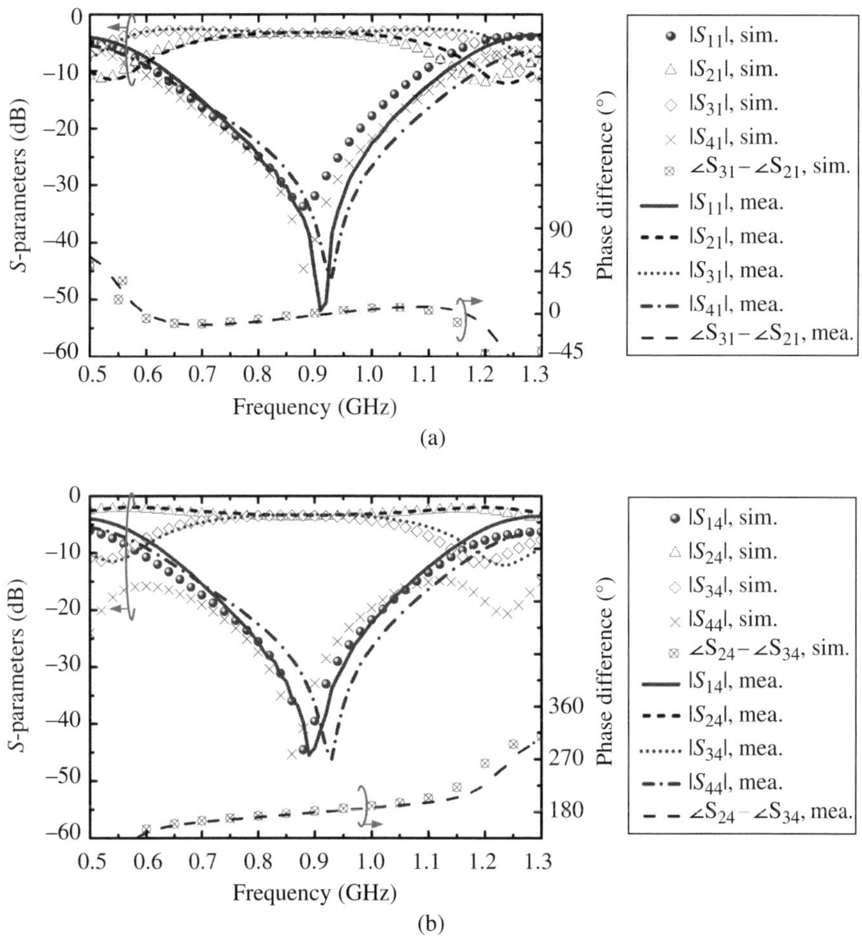

Figure 6.12 *S*-parameters and output phase differences of the rat-race coupler when the (a) sum (Σ) port and (b) difference (Δ) port serves as the input

coefficient is lower than −45 dB while the isolation is better than 40 dB in both cases. As the sum port (port 1) is excited, the transmission coefficients ($|S_{21}|$, $|S_{31}|$) are −3.3 and −3.2 dB, and the associated phase difference is 2.9°. Meanwhile, they become −3.4 and −3.2 dB ($|S_{24}|$, $|S_{34}|$) when the difference (Δ) port serves as the excitation. The outputs are almost out of phase by 180°. The operating bandwidth is 22–24%.

6.4 Broadside Directional Coupler

6.4.1 Design Principle

The operational principle of coupled transmission lines and their applications to directional couplers have been introduced and analyzed in Sec. 2.4. In this section, a similar design procedure is applied to develop a broadside directional coupler using periodic synthesized transmission lines.

Figure 6.13 depicts the equivalent lumped model of a pair of coupled periodic synthesized lines. It is a four-port network and p is the periodicity. Each stand-alone synthesized line comprises n unit cells, and the pairs of cells in the two lines are electrically and magnetically coupled to one another via L_m and C_m. Since the structure is fully symmetric with respect to the central plane, the even-odd mode analysis can be applied to convert the network to half circuits as indicated in Fig. 6.14. The even- and odd-mode characteristic impedances and electrical lengths of the coupling section are therefore

$$Z_{0e} = \sqrt{\frac{L_e}{C_e}} \tag{6.10}$$

$$Z_{0o} = \sqrt{\frac{L_o}{C_o}} \tag{6.11}$$

$$\theta_e = n\beta_e p = n\omega\sqrt{L_e C_e} \tag{6.12}$$

$$\theta_o = n\beta_o p = n\omega\sqrt{L_o C_o}, \tag{6.13}$$

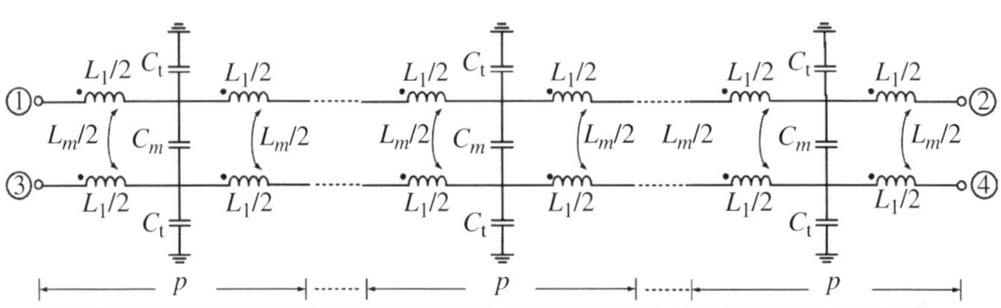

Figure 6.13 Equivalent lumped model of a pair of coupled periodic synthesized lines

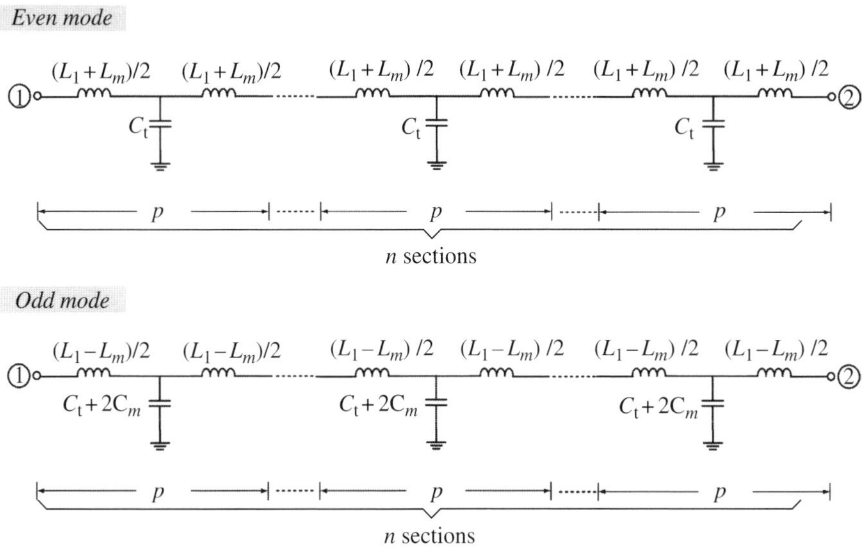

Figure 6.14 Even- and odd-mode half circuits of the coupled periodic synthesized lines

where,

$$L_e = L_1 + L_m \tag{6.14}$$

$$C_e = C_t \tag{6.15}$$

$$L_o = L_1 - L_m \tag{6.16}$$

$$C_o = C_t + 2C_m. \tag{6.17}$$

From (6.10)–(6.17) and under the TEM assumption that the electrical lengths of the two half circuits are the same at the center frequency

$$\theta_e = \theta_o = \left.\frac{\pi}{2}\right|_{\omega=\omega_0}, \tag{6.18}$$

the lumped values (L_1, C_t, L_m, and C_m) can be solved in terms of Z_{0e}, Z_{0o}, and ω_0 as

$$L_1 = \frac{L_e + L_o}{2} = \frac{\pi}{2n\omega_0}\left(\frac{Z_{0e} + Z_{0o}}{2}\right) \tag{6.19}$$

$$L_m = \frac{L_e - L_o}{2} = \frac{\pi}{2n\omega_0}\left(\frac{Z_{0e} - Z_{0o}}{2}\right) \tag{6.20}$$

$$C_t = C_e = \frac{\pi}{2n\omega_0}\left(\frac{1}{Z_{0e}}\right) \tag{6.21}$$

$$C_m = \frac{C_o - C_e}{2} = \frac{\pi}{4n\omega_0}\left(\frac{Z_{0e} - Z_{0o}}{Z_{0e}Z_{0o}}\right). \tag{6.22}$$

For given coupling strength in a particular design, say, a coupling coefficient of 3 dB or $k = 0.707$ in (2.21), the even-/odd-mode characteristic impedances can be first solved using (2.20) and (2.21) as

$$Z_{0e} = Z_0\sqrt{\frac{1+k}{1-k}} \tag{6.23}$$

$$Z_{0o} = Z_0\sqrt{\frac{1-k}{1+k}}. \tag{6.24}$$

The lumped values to fulfill the design are then fixed by (6.19)–(6.22) once the number of sections n is specified.

6.4.2 Circuit Realization

As an illustration, a 3-dB directional coupler in the 50-Ω system is developed using the periodic unit cell in Fig. 6.1. The targeted even- and odd-mode characteristic impedances are 120.7 and 20.7 Ω, and n is set to 12. By solving the equations, the lumped values are $L_l = 1.62$ nH, $L_m = 1.16$ nH, $C_t = 0.19$ pF, and $C_m = 0.48$ pF if the center frequency is also set to 900 MHz. The layout of the unit cell to complete the coupler is shown in Fig. 6.15. In this design, the center patch is removed and the capacitive pad in each quadrant is confined to a small area since the required self capacitance is very small as 0.19 pF. The pads are connected to the cross junction using 0.15-mm thin lines.

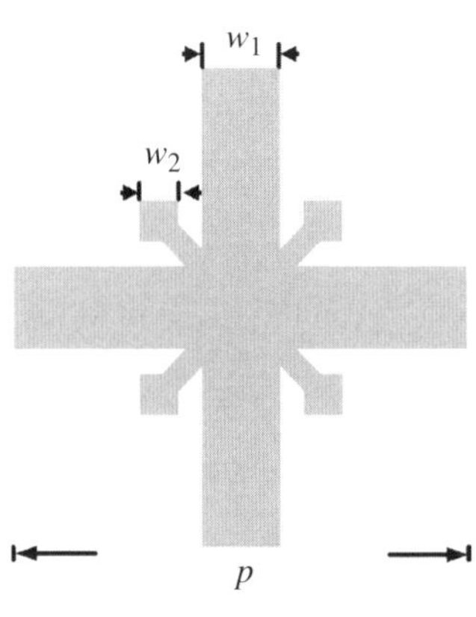

Figure 6.15 Layout of the unit cell in the coupled periodic synthesized lines

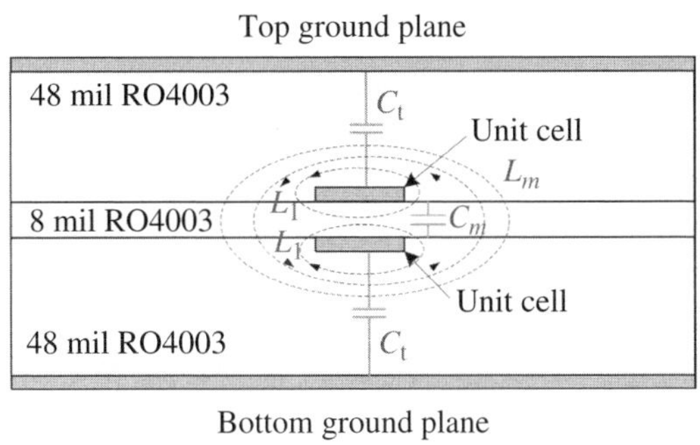

Figure 6.16 Cross-sectional profile of the broadside coupler using periodic cells

The mutual capacitance as well as mutual inductance, on the other side, is quite large; it suggests that the two lines must be placed in close proximity. Figure 6.16 shows the cross-sectional profile of the coupling structure. It is a broadside coupler using multilayered PCB technology, and the substrate is Rogers RO4003C. The unit cells are separated by a thin substrate layer whose thickness is 8 mil to facilitate tight coupling. The thin layer is then sandwiched by two thick layers of 48 mil to keep the self capacitance between the unit cell and ground plane a small value to meet the design specification.

Figure 6.17(a) and (b) analyzes the even-/odd-mode characteristic impedances and effective dielectric constants versus geometric parameters w_1 and w_2. The corresponding coupling coefficient is plotted in Fig. 6.17(c). Referring to Fig. 6.17(a), the characteristic impedances (and hence the coupling coefficient) are mainly a function of the strip width w_1. It determines the mutual inductance as well as a major portion of the mutual capacitance at the same time. The area of the capacitive pad (w_2) only has a minor effect on the impedances. By increasing w_2 or equivalently the C_t, the impedances of both modes decrease slightly. The effective dielectric constants in Fig. 6.17(b) reveal a different trend. While the even-mode parameters are also determined by the strip width w_1, the odd-mode ones are controlled by both parameters, especially when w_1 is small. It is because that when the strips become narrow, the capacitive pads contribute more, on a relative basis, to the mutual capacitance. The odd-mode propagation characteristics are affected accordingly.

Using the design graphs, the geometric parameters of the unit cells are determined as $w_1 = 0.5$ mm and $w_2 = 0.25$ mm. The periodicity p is 2.9 mm. The layout of the broadside coupler is shown in Fig. 6.18(a), and a photograph of the fabricated sample is shown in Fig. 6.18(b). The multilayered structure is realized by stacking and fixing the layers using plastic screws. The simulated and measured S-parameters and phase differences at outputs are summarized and compared in Fig. 6.19(a) and (b), respectively. At 900 MHz, the reflection coefficient $|S_{11}|$ is −17 dB while the isolation $|S_{41}|$ is 13.8 dB. The transmission coefficients at the thru and coupled ports ($|S_{21}|$, $|S_{31}|$) are −3.45 and −3.2 dB. In addition to fabrication tolerance, the fairly low isolation and directivity is also a result of the unequal phase velocities in both modes, as suggested in Fig. 6.17(b). The phase response, on the other side, remains very good over a wide bandwidth. At the center frequency, the phase difference is 86.5°. The fractional bandwidth of the coupler is similar to its conventional counterpart as 50%. When compared with a broadside coupler using uniform coupled striplines, the reduction in coupling length reaches 87%.

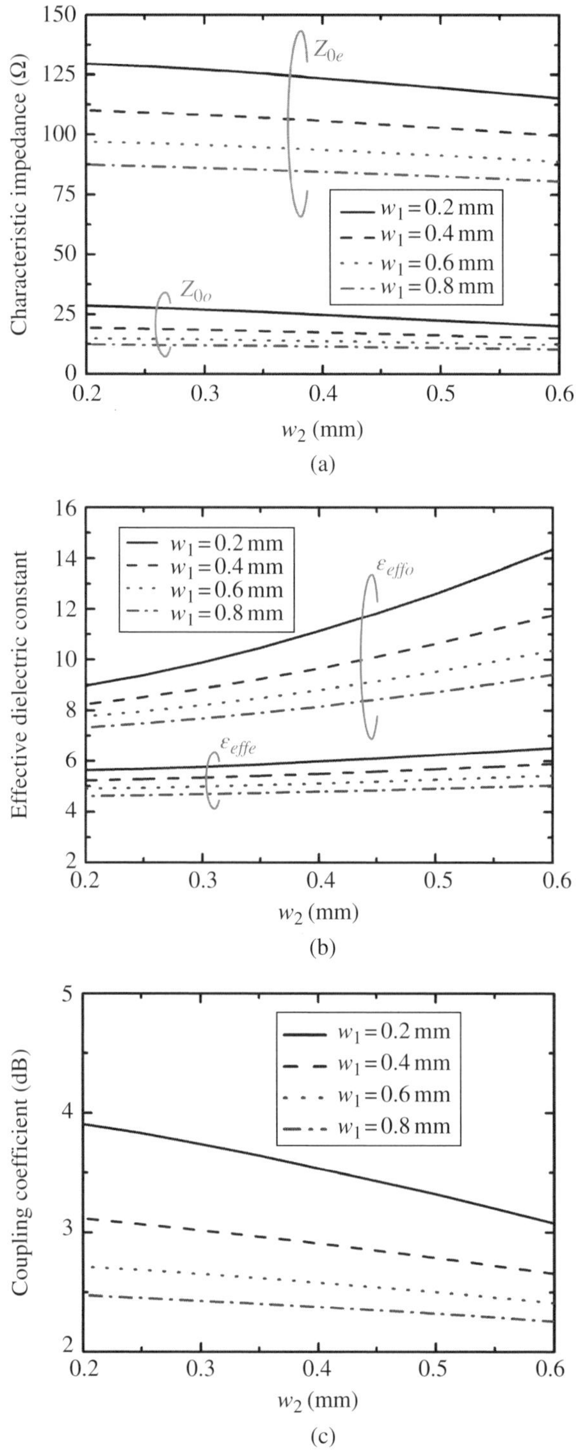

Figure 6.17 Design graphs of the broadside coupler using periodic synthesized lines

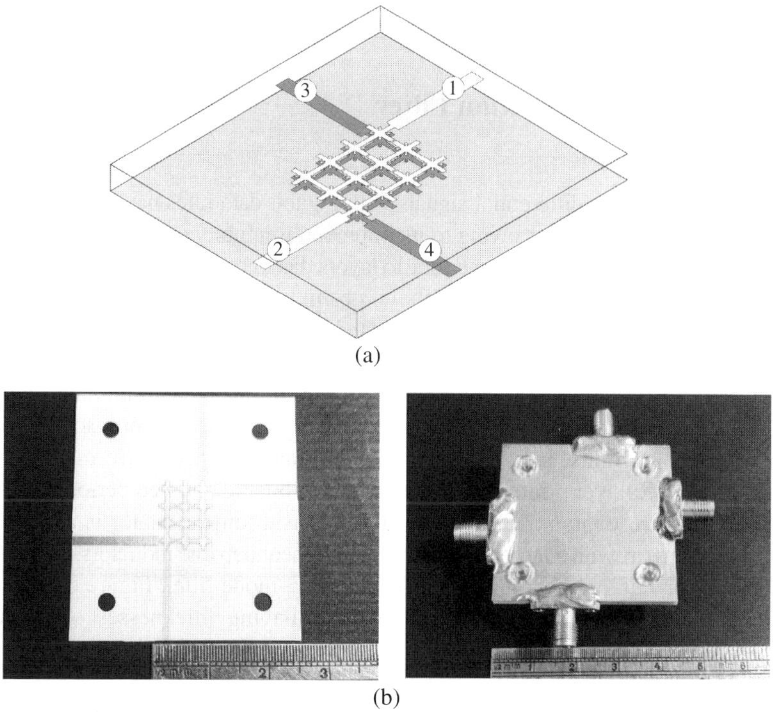

Figure 6.18 (a) Layout and (b) photograph of the fabricated sample of the broadside coupler realized by periodic synthesized lines

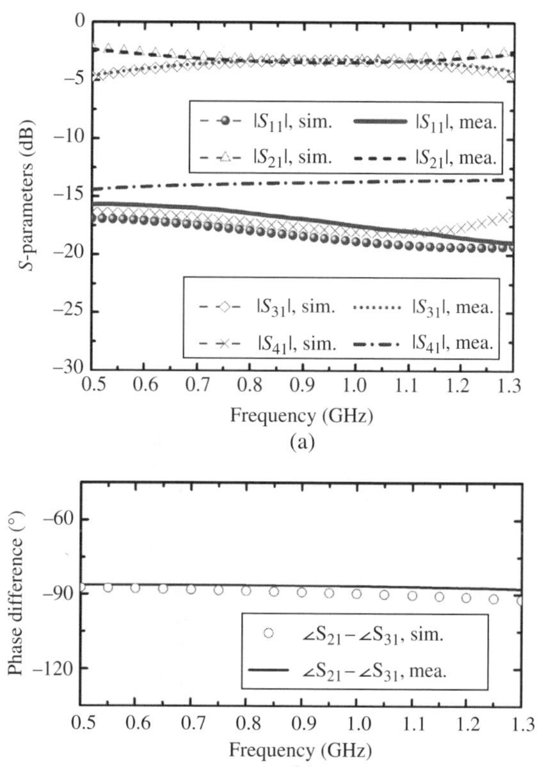

Figure 6.19 (a) *S*-parameters and (b) output phase differences of the broadside coupler

6.5 Common-Mode Rejection Filter

6.5.1 Design Principle

As mentioned in Sec. 5.4.3, differential signals with gigabit data rates have been widely used in modern digital computing systems owing to its improved immunity against noises and EM interferences. In practical designs, the unbalance due to layout discontinuity inevitably leads to unwanted common-mode noise/radiation. To tackle the problem, common-mode rejection filters have become an indispensable component to be integrated with high-speed differential pairs [26–30]. Lumped elements with common-mode choke resonators [26–28] and transmission-line pairs with transmission zeros (TZs) [29, 30] are common approaches for realizing a common-mode filter.

Periodic synthesized transmission lines can be applied to realize a compact common-mode filter with a good compromise between the circuit size and electrical performance. The design is developed on a three-layer printed circuit board (PCB) using coupled periodic unit cells and modified mushroom resonators [31]. The modified mushroom resonator is a variation of its conventional counterpart with a larger inductance using meander-line inductors. Figure 6.20(a–c) shows the circuit layout and photograph of the common-mode filter. Figure 6.21 is the cross-sectional profile. Two Rogers RO4003C substrates having thicknesses of 8 and 60 mil,

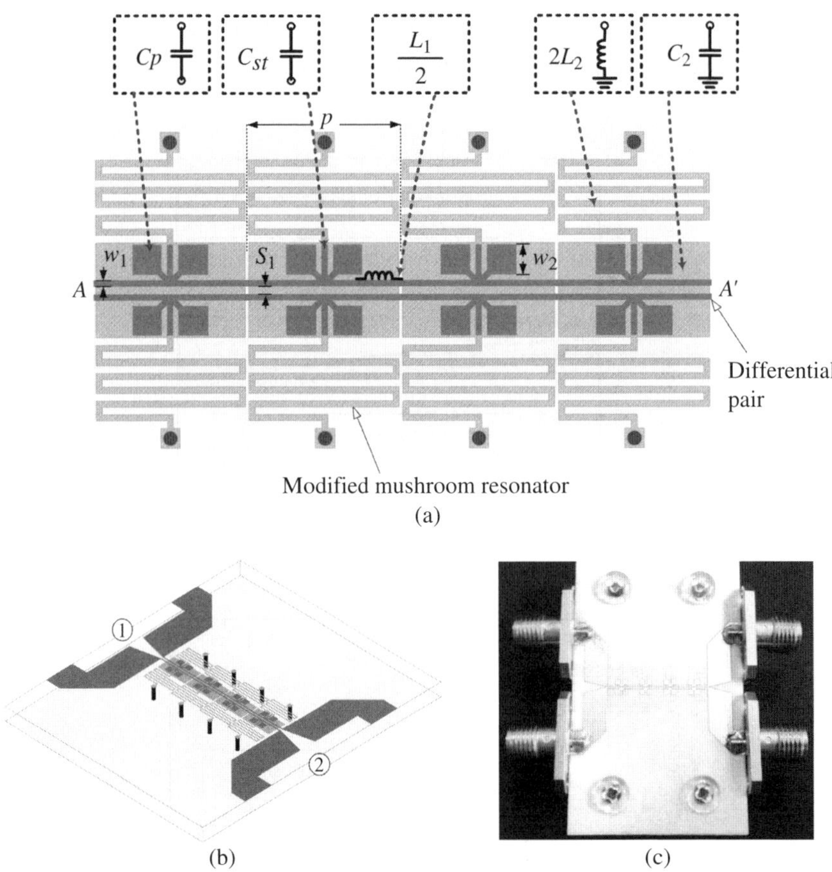

Figure 6.20 (a) Top view, (b) three-dimensional view, and (c) photograph of the common-mode filter

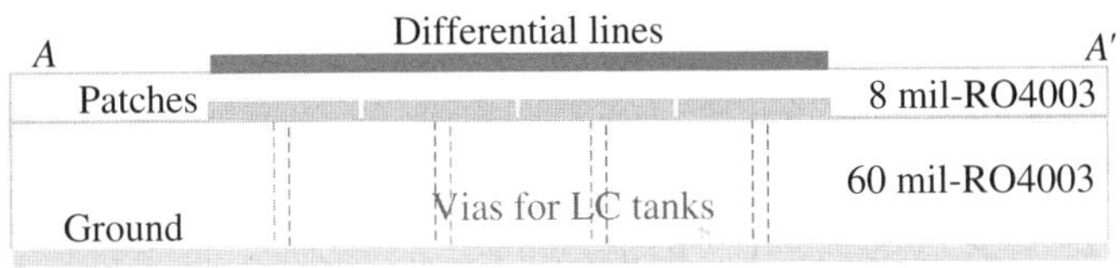

Figure 6.21 Cross-sectional profile of the common-mode filter. Source: Su 2015 [31]. Reproduced with permission of IEEE

respectively, are stacked up together. A standard double-layer lithographic etching process is first performed, followed by an alignment procedure with plastic screws.

The common-mode filter comprises a periodic connection of four identical unit cells, each one having a length (periodicity) of p. On the top layout of the thin substrate, the unit cell in Fig. 6.1 is bisected to form a differential pair. Each cell comprises a thin inductive line, an open stub, and two capacitive pads. The mushroom resonator is developed on the thick substrate and serves as the common-mode choke resonator in the shunt path. Two meander-line inductors are connected to the edges of the capacitive patch to allocate the resonance frequency of the tank at the targeted common-mode rejection band. Without a patterned ground plane, the parasitic radiation and unwanted EM interference can be alleviated, and the power integrity is retained.

Since the unit cell is much shorter than the wavelength, it can be represented by a lumped equivalent model as shown in Fig. 6.22(a). Here, C_p and C_{st} represent the capacitances between the pads/stub on the top layer and the capacitive patch of the mushroom. The L_2 and C_2 are the total inductance and capacitance of the mushroom resonator. L_1 and L_m are the self and mutual inductances of the inductive differential pair, and C_m is the mutual capacitance between the lines.

Due to the geometrical symmetry, the even-odd mode analysis can be applied and the half circuits are shown in Fig. 6.22(b). In the odd mode, the unit cell is equivalent to a simple T-network, inherently behaving like a transmission line with low-pass responses. With the assumption of infinite periodicity, the Bloch impedance and dispersion relationship are derived as

$$Z_{B,odd} = \sqrt{\frac{2Z_{odd}}{Y_{odd}}} = \sqrt{\frac{L_1 - L_m}{C_1 + 2C_m}} \tag{6.25}$$

$$\beta_{odd} p = \omega\sqrt{(L_1 - L_m)(C_1 + 2C_m)} \tag{6.26}$$

$$C_1 \triangleq 2C_p + C_{st}. \tag{6.27}$$

The operating band of the odd mode is from DC to f_c,

$$f_c = 1/\pi\sqrt{(L_1 - L_m)(C_1 + 2C_m)}. \tag{6.28}$$

In the meantime, when the unit cell is operated in the even mode, the plane of symmetry is virtually open-circuited such that the parallel LC resonator from the mushroom is in series

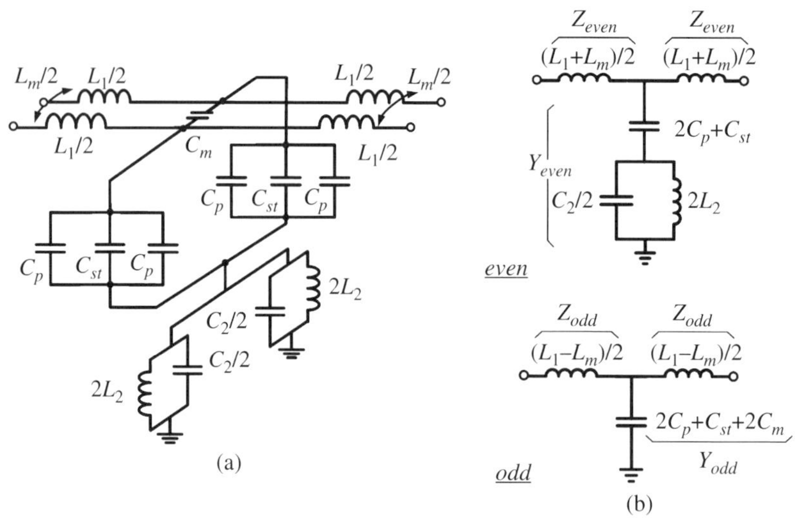

Figure 6.22 (a) Complete and (b) even-/odd-mode equivalent circuit models of the unit cell in the common-mode filter. Source: Su 2015 [31]. Reproduced with permission of IEEE

with the shunt capacitance C_1. By also assuming that the cells are cascaded with infinite periodicity, the dispersion relationship of the Bloch wave in the even mode is

$$\cosh(\gamma_{even} p) = 1 - \omega^2 (L_1 + L_m) C_1 \left(1 - \frac{\omega^2}{\omega_o^2}\right) \Big/ 2\left(1 - \frac{\omega^2}{\omega_z^2}\right). \tag{6.29}$$

In (6.29),

$$\omega_o = 2\pi f_o = 1 \big/ \sqrt{L_2 C_2} \tag{6.30}$$

and

$$\omega_z = 2\pi f_z = 1 \big/ \sqrt{L_2 (2C_1 + C_2)} \tag{6.31}$$

are the resonance frequencies of the LC tank ($2L_2$ // $C_2/2$) and the whole shunt network (Y_{even}), respectively. $\gamma_{even} = \alpha_{even} + j\,\beta_{even}$ is the complex propagation constant. According to (6.29), in the even mode with infinite periodicity, there is a bandgap ranging from f_z to f_o, where the propagation constant is purely real ($\alpha_{even} \neq 0$, $\beta_{even} = 0$) and represents an evanescent mode. It prevents the common mode signals from passing through the differential line pair, and creates the desired band-notched phenomenon.

Equations (6.25), (6.30), and (6.31) are the key formulae of the common-mode filter. As there are only three equations but six unknowns (L_1, L_2, L_m, C_1, C_2, C_m), one has three degrees of freedom that should be selected in advance. Here, the line width and spacing of the differential pair (w_1 and s_1) are selected equal to the minimum attainable resolution that the PCB process was allowed. This, in turn, fixes the parameters irrelevant to the geometry of the mushroom resonator, including L_1, L_m, and C_m. With L_1, L_m, and C_m specified, the capacitance C_1 is determined using (6.25) for a given odd-mode characteristic impedance, say 50 Ω, in the passband. The remaining two parameters associated with the mushroom resonator are solved from the simultaneous equations (6.30) and (6.31) once the desired stopband (f_z, f_o) is specified.

Here, the parasitic coupling between unit cells is ignored, but a more accurate prediction can be found in [26, 28] using an analytical model with the four-port even-mode *Z*-matrix.

6.5.2 Circuit Realization

Following the design guideline, the common-mode filter is experimentally verified. In order to achieve an overall compact size, the periodicity length p is set to 2.8 mm. With $w_1 = s_1 = 0.15$ mm, the parameters (L_l, L_m, C_m) are extracted as $L_l = 3.11$ nH, $L_m = 1.7$ nH, and $C_m = 0.029$ pF. By setting $Z_{B,\,odd} = 50\ \Omega$, $f_z = 1.8$ GHz and $f_o = 3.0$ GHz, the remaining three parameters are solved as $L_2 = 4.16$ nH, $C_1 = 0.586$ pF, and $C_2 = 0.68$ pF. The lumped values are converted into physical footprints, with the capacitance C_1 replaced by two capacitive pads and an open stub. Finally, the parts are integrated and a fine-tuning process is performed to account for the parasitic coupling between the elements.

Figure 6.23(a) depicts the full-wave simulated and model-calculated dispersion diagrams of the unit cell in the even and odd modes. The odd-mode phase constant is a linear function of

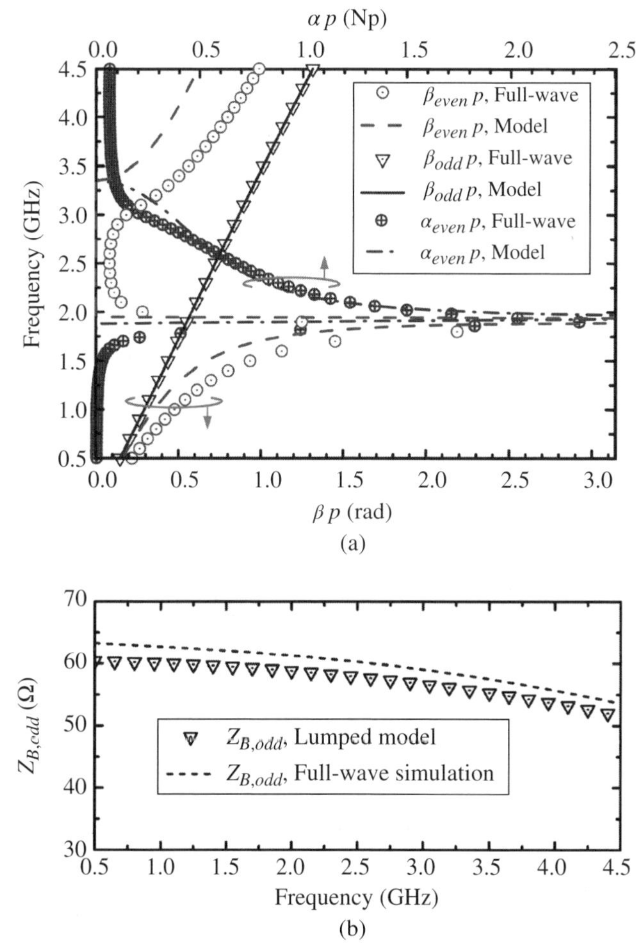

Figure 6.23 (a) Even- and odd-mode dispersion diagram and (b) odd-mode Bloch impedance of the unit cell

frequency. The simulated Bloch impedance, as shown in Fig. 6.23(b), is 55 Ω over the considered band. The even-mode complex propagation constant, meanwhile, indicates a bandgap between 1.9 and 3.3 GHz. Its Bloch impedance is imaginary in the stopband.

Four unit cells are cascaded as the common-mode filter. The overall occupied size is only $0.113 \times 0.226\,\lambda_g^2$. For the PCB design, the unit cells are connected one-dimensionally since the mushroom resonators may overlap with each other if bends are included. Nevertheless, an integrated fabricated process such as the LTCC technology may alleviate the problem and hence facilitate real two-dimensional routing. The common-mode (CM) and differential-mode (DM) S-parameters are summarized in Fig. 6.24. From 1.8 to 3.7 GHz, the measured $|S_{cc21}|$ and $|S_{dd11}|$ are both below −15 dB, while the 3-dB cut-off frequency of the differential mode ($|S_{dd21}|$) is 8.5 GHz. The common-mode rejection is higher than 40 dB at 2.5 GHz. The differential-mode transmission loss is 0.5 dB at the same frequency. The fractional bandwidth with $|S_{cc21}| < -15$ dB is more than 70%. Figure 6.25 compares the differential-signal eye diagram of the proposed common-mode filter to that of a reference board. The reference board is realized by a direct connection of the I/O ports by a pair of connecting lines of the same length. The eye diagrams are achieved by importing the measured frequency-domain S-parameters into the circuit emulator to perform a channel simulation with a 5-Gbps PRBS differential signal. Comparable eye widths and heights are observed, indicating almost identical signal integrity for both cases.

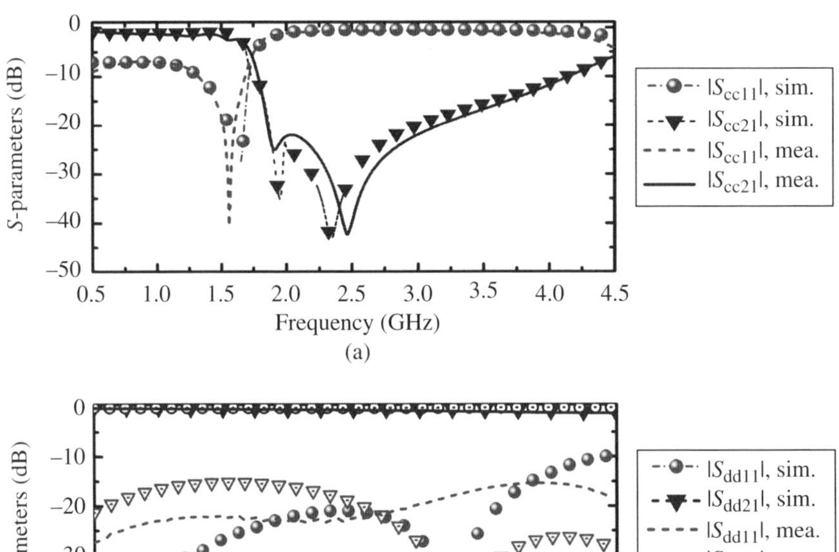

(a)

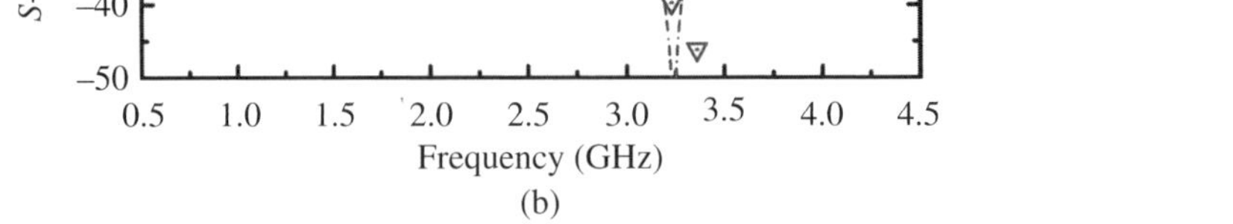

(b)

Figure 6.24 (a) Common-mode S-parameters and (b) differential-mode S-parameters of the common-mode filter

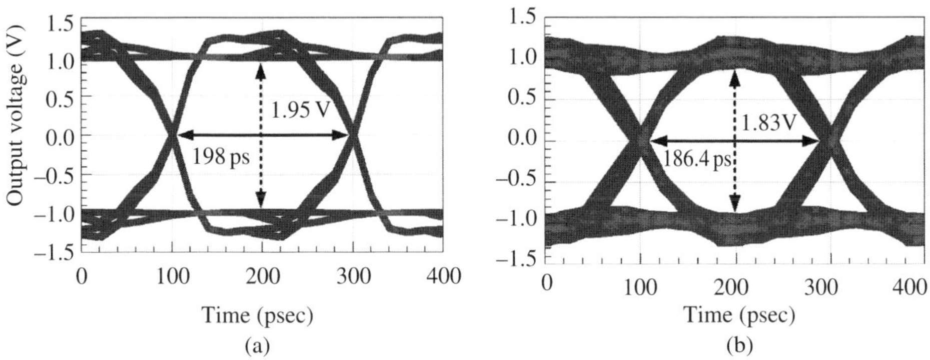

Figure 6.25 Eye diagrams for the (a) reference board and (b) proposed common-mode filter. Source: Su 2015 [31]. Reproduced with permission of IEEE

6.6 On-Chip Implementation

6.6.1 Unit Cells and Quarter-Wavelength Lines

The periodic synthesized transmission lines can be implemented as on-chip designs by using the integrated passive device (IPD) process, as well [32]. The on-chip version may fully exploit the two-dimensional layout capability by routing the unit cells around a number of circuit blocks to achieve a very efficient usage of the chip area.

Figure 6.26 shows the circuit layout and corresponding lumped model of the unit cell for realizing the on-chip periodic synthesized transmission line. It is a modified version of the unit cell in Fig. 6.1 to take advantage of the multilayered IPD process. A 35-Ω line is given as an example. The unit cell is developed on the silicon-based IPD substrate. The readers may refer to Sec. 5.2 for more information regarding the IPD technology.

The unit cell is square in shape. The side of the square, or the periodicity, is p (hereafter set as 260 μm). An asterisk-shaped conducting strip on the M3 layer is centered at the cell and a square patch is overlapped with it. The asterisk can be further decomposed into a x-shaped and a +-shaped. Two branches of the +-shaped serves as the input/output ports to adjacent cells, and hence provides the series inductance (L_1) of the line. The remaining two branches, together with the x-shaped, fulfill the capacitive part of the unit cell (C_2 and C_3). The co-centered square patch is also capacitive as C_1. Similar to the center patch in Fig. 6.1, it is responsible for adjusting the impedance level of the line. The traces on the M1 layer serve as the reference ground; they are an image of the layout on the M3 layer to form MIM capacitors between the layers. A ground ring is added to the peripheral of the square. The stubs on the M1 layer are connected to the ground ring to keep equal potential for proper operation. Four M3-to-M1 vias are added to improve the isolation between the unit cell and surroundings.

The synthesis procedure of the on-chip design is very similar to that using the PCB technology in Fig. 6.1. They share the same lumped model as shown in Fig. 6.2. For a given impedance level and electrical length, the required total inductance (L_1) and capacitance (C_t) are first solved using the simultaneous equations (6.7) and (6.9). The length of the line inductance (l_1) is then selected to fix the pair of series inductance ($L_1/2$). Since the periodicity has been set as p, l_1 also determines the size of the capacitive patch ($l_2 = p - 2l_1$) and its associated

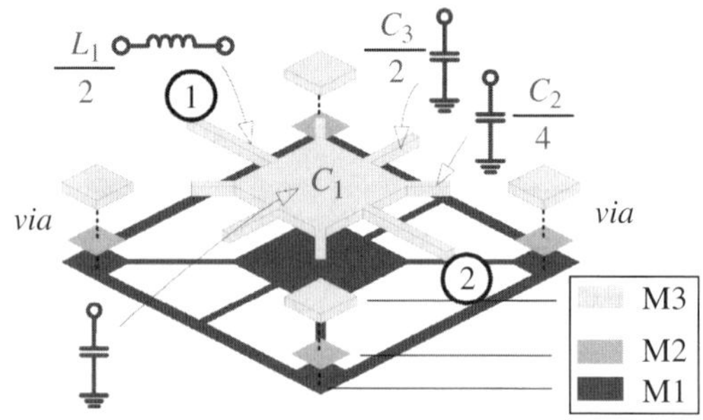

Figure 6.26 Three-dimensional view of the unit cell for realizing on-chip periodic synthesized transmission lines.

capacitance (C_1). Finally, the lengths of the stubs (l_3 and l_4) and its overlapped region with the M1 layer are adjusted for proper values of C_2 and C_3 to fulfill the given total capacitance C_t, and hence complete the design. Since the only constraint imposed on the design is the total capacitance C_t, the combination of l_3 and l_4 is relatively arbitrary. In this design, we simply keep the + and ×-shaped strips almost equal in length for symmetry reasons.

For an on-chip unit cell with a higher impedance, the metallic stubs on the M1 layer could be partially or completely removed, transforming the strips on the M3 layer into open stubs with a smaller capacitance. As an illustration, the top views of a 35-, 50-, 70-Ω unit cells are shown in Fig. 6.27(a)–(c). The series inductance and total capacitance are determined as (L_1=0.132 nH, C_t=0.102 pF) for the 35-Ω cell, (L_1=0.155 nH, C_t=0.059 pF) for the 50-Ω cell, and (L_1=0.204 nH, C_t=0.040 pF) for the 70-Ω cell. The per-unit-cell phase delays are 7.3°, 6.0°, and 5.7°, respectively. For the 50-Ω cell, the stubs on the M1 layer associated with the + strip (l_4) are removed, and the area of the center pad on the same layer is shrunk to decrease the MIM capacitance C_1. At 5.5 GHz, the total capacitance drops accordingly and hence increases the Bloch impedance of the unit cell. For the 70-Ω cell, the metallic patterns on the M1 layer are completely removed to minimize the shunt capacitance. A unit cell with impedance higher than 70 Ω is also possible, but the limitation is around 90 Ω, at which the synthesized line becomes no more than a common meander line by removing all stubs from the inductive host line. On the contrary, the lower bound of the realizable impedance is around 20–25 Ω, at which the unit cell is filled out by the capacitive pad C_1.

By connecting the unit cells in cascade, quarter-wavelength lines are realized and shown in Fig. 6.28(a)–(c). The numbers of cells in the 35-, 50-, and 70-Ω 90° lines are 12, 16, and 18, respectively. All lines are routed two dimensionally, and a chip photo of the 70-Ω 90° line is shown in Fig. 6.28(d). Figure 6.29 summarizes the simulated and measured characteristic impedances and electrical lengths of the lines. The transmission loss is around 0.4–0.6 dB, while the electrical length is 85–90° at the center frequency of 5.5 GHz. When calculating the direct product of the phase delay of a single unit cell by the number of cells, the readers may note that there is a discrepancy between the targeted and measured electrical lengths. The discrepancy is a result of the unaccounted parasitic inductive coupling between the cells, which decreases the actual electrical length of the cells in the quarter-wavelength line. To tackle the problem, a few extra unit cells or short conventional transmission lines can be added to compensate for the slight difference between the targeted and actual electrical lengths of the synthesized line.

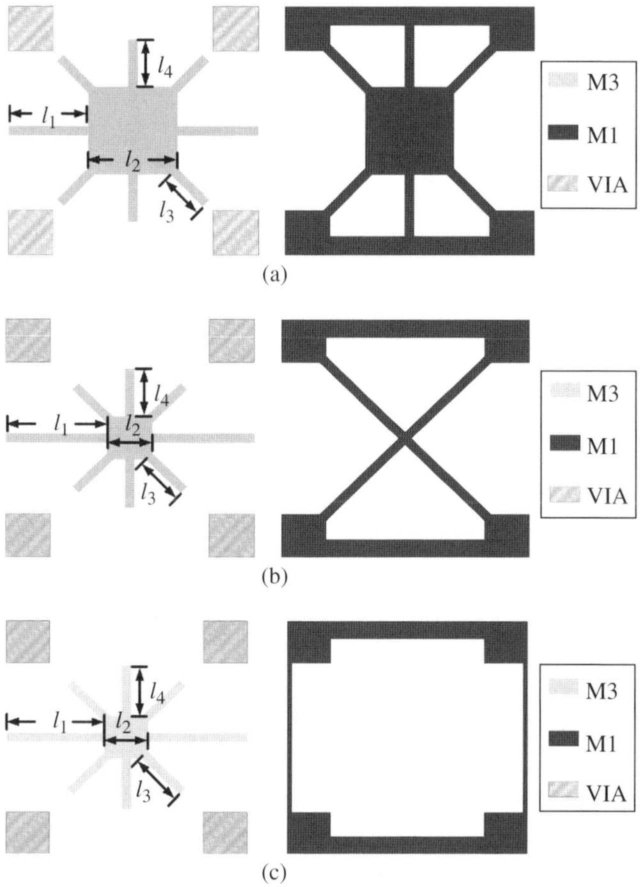

Figure 6.27 Top views of the on-chip unit cells: (a) 35 Ω, (b) 50 Ω, and (c) 70 Ω

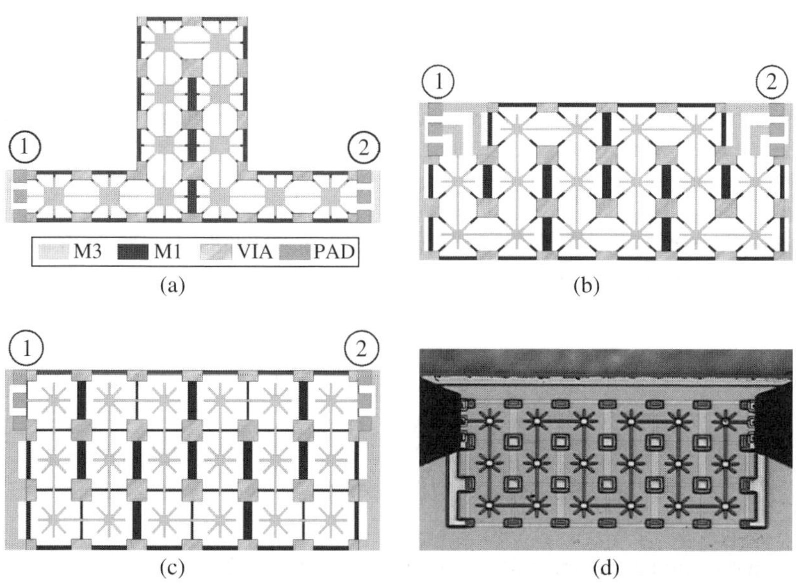

Figure 6.28 Layouts of quarter-wavelength lines on the IPD process: (a) 35 Ω, (b) 50 Ω, and (c) 70 Ω; (d) chip photo of the 70 Ω 90° line

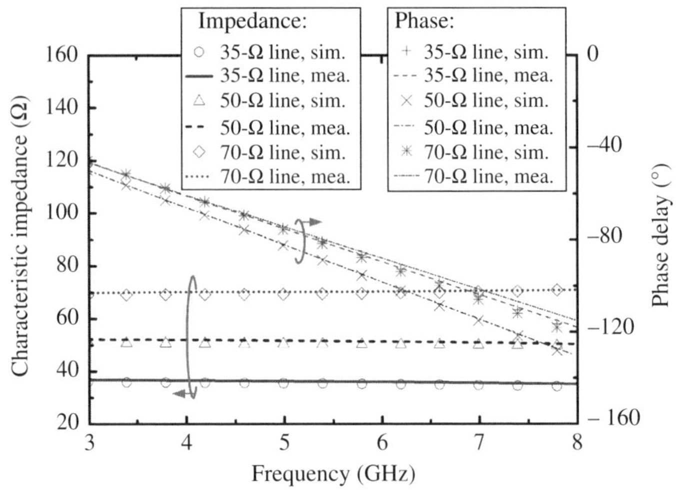

Figure 6.29 Characteristic impedances and electrical lengths of the quarter-wavelength lines using periodic unit cells on the IPD process

6.6.2 Circuit Implementations and Compensation

The quarter-wavelength lines are directly integrated as a Wilkinson power divider, branch-line coupler, and rat-race coupler. The chip photos are shown in Fig. 6.30(a)–(c). When compared with their conventional counterparts using the same process, the size reduction ratios are 86, 93.9, and 95.8%, respectively. Figure 6.31 plots the circuit response of the rat-race coupler when excited at the difference port. It has the longest transmission path and hence highest power dissipation. The transmission loss is around 1.2 dB, while the amplitude and phase balances are good over a fractional bandwidth of 20%.

The readers may also note that in the two-dimensional routing, a number of bends is inevitably required to fulfill the layout. The bend alters the current distribution of the unit cell and hence its electrical properties. For a conventional uniform microstrip line, a chamfered or notched structure is introduced to compensate for the excess capacitance due to the bend. The complementary-conducting-strip (CCS) lines generally require an equal number of right and left bends to neutralize the discontinuity effects. An alternative approach to treat the bends is to view them as independent corner cells [32]. The corner cell can be developed as a non-periodic lumped network whose characteristic impedance and electrical length are described using a different set of equations. Again, by equating the *ABCD* matrix of the corner cell to that of a uniform transmission line of (Z_c, θ_c), the design equations are

$$Z_c = \sqrt{\frac{B}{C}} = \sqrt{\frac{L_1 - \omega^2 L_1^2 C_t / 2}{C_t}} \tag{6.32}$$

$$\theta_c = \cos^{-1}\left(1 - \omega^2 L_1 C_t / 2\right). \tag{6.33}$$

More information and experimental results can be found in [32].

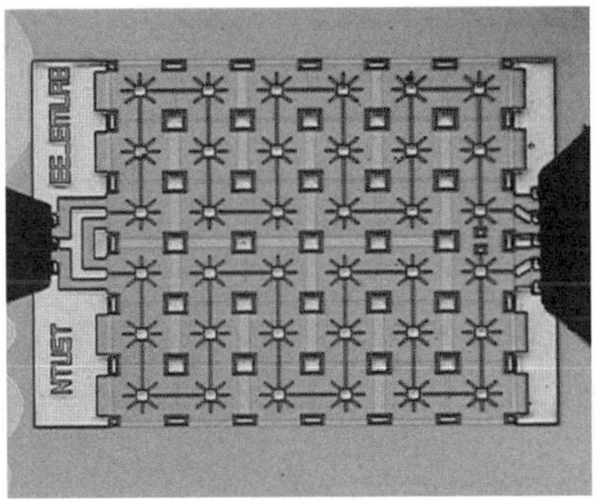

(a)

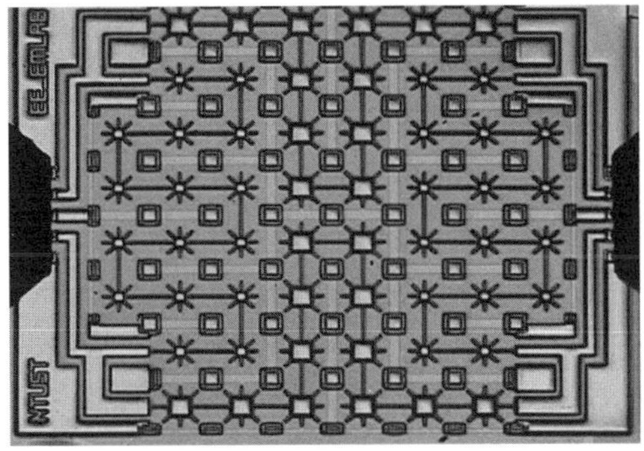

(b)

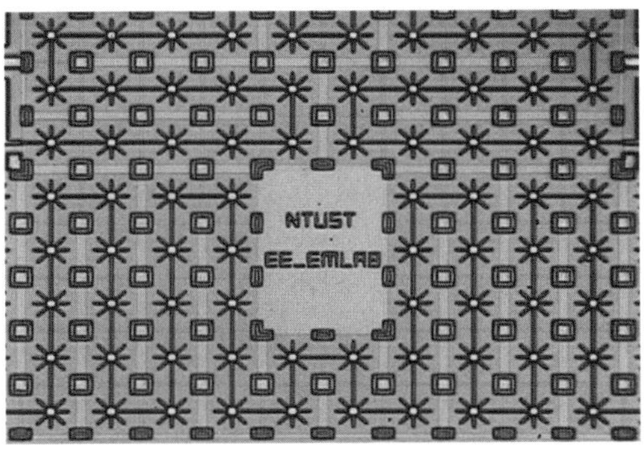

(c)

Figure 6.30 Chip photos of the (a) Wilkinson power divider, (b) branch-line coupler, and (c) rat-race coupler using periodic synthesized lines on the IPD process

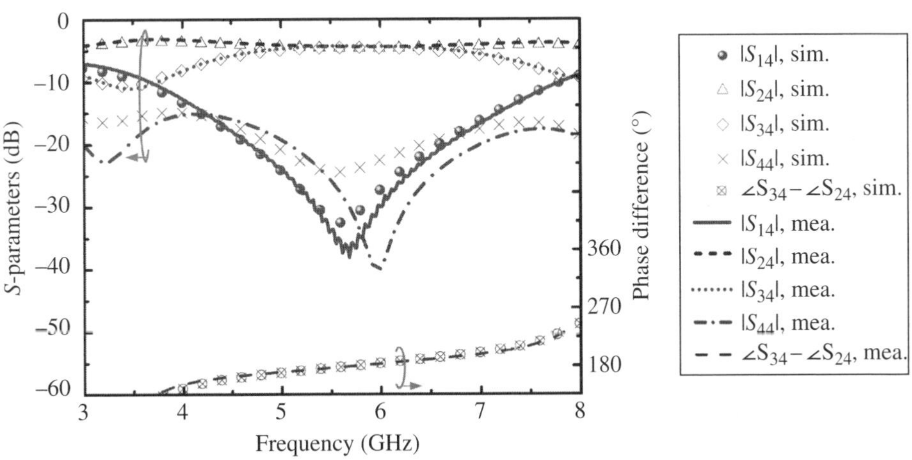

Figure 6.31 *S*-parameters and output phase differences of the on-chip rat-race coupler when the difference (Δ) port is excited

Last but not the least, the readers should keep in mind that the layout does not always have to be in rectangular shape; it could be conformal to an arbitrary outline profile when integrated in a real chip design.

References

[1] K.-O. Sun, S.-J. Ho, C.-C. Yen, and D. V. C. Weide, "A compact branch-line coupler using discontinuous microstrip lines," *IEEE Microw. Wireless Compon. Lett.*, vol. 15, no. 8, pp. 519–520, Aug. 2005.

[2] Z.-Y. Liu and R. M. Weikle, "A compact quadrature coupler based on coupled artificial transmission lines," *IEEE Microw. Wireless Compon. Lett.*, vol. 15, no. 12, pp. 889–891, Dec. 2005.

[3] C.-C. Chen and C.-K. C. Tzuang, "Synthetic quasi-TEM meandered transmission lines for compacted microwave integrated circuits," *IEEE Trans. Microw. Theory Techn.*, vol. 52, no. 6, pp. 1637–1647, Jun. 2004.

[4] M.-J. Chiang, H.-S. Wu, and C.-K. C. Tzuang, "Design of synthetic quasi-TEM transmission line for CMOS compact integrated circuit," *IEEE Trans. Microw. Theory Techn.*, vol. 55, no. 12, pp. 2512–2520, Dec. 2007.

[5] C.-C. Chen, J.-J. Cin, S.-H. Wang, C.-C. Lin, and C.-K. C. Tzuang, "A novel miniaturized wideband Wilkinson power divider employing two-dimensional transmission line," in *Proc. IEEE Int. VLSI Design, Automation and Test Symp.*, 2008, pp. 212–215.

[6] M.-J. Chiang, H.-S. Wu, M.-L. Lee, and C.-K. C. Tzuang, "Design of compact Ka-band monolithic branch-line coupler on silicon substrate," in *Proc. Asia-Pacific Microw. Conf.*, 2009, pp. 2124–2127.

[7] P.-L. Chi and T. Itoh, "Miniaturized dual-band directional couplers using composite right/left-handed transmission structures and their applications in beam pattern diversity systems," *IEEE Trans. Microw. Theory Techn.*, vol. 57, no. 5, pp. 1207–1215, May. 2009.

[8] C.-W. Wang, T.-G. Ma, and C.-F. Yang, "A new planar artificial transmission line and its applications to miniaturized Butler matrix," *IEEE Trans. Microw. Theory Techn.*, vol. 55, no. 12, pp. 2792–2801, Dec. 2007.

[9] C.-H. Lai, C.-Y. Shiau, and T.-G. Ma, "Microwave three-channel selector using tri-mode synthesized transmission lines," *IEEE Trans. Microw. Theory Techn.*, vol. 61, no. 10, pp. 3529–3540, Oct. 2013.

[10] I. Haroun, T.-Y. Lin, D.-C. Chang, and C. Plett, "A reduced-size, low-loss 57–86 GHz IPD-based power divider using loaded modified CPW transmission lines," in *Proc. Asia-Pacific Microw. Conf.*, 2012, pp. 1202–1204.

[11] Y.-J. Li, M.-J. Xing, Z.-M. Zhu, and Y.-T. Yang, "Novel compact compass navigation system (CNS) power divider," in *Proc. Int. Conf. Electronic Packaging Technology & High Density Packaging*, 2010, pp. 710–713.

[12] H.-T. Kim, K. Liu, R. C. Frye, Y.-T. Lee, G. Kim, and B. Ahn, "Design of compact power divider using integrated passive device (IPD) technology," in *Proc. Electronic Components and Technology Conf.*, 2009, pp. 1894–1899.

[13] M. Ercoli, D. Dragomirescu, and R. Plana, "An extremely miniaturized ultra wide band 10–67 GHz power splitter in 65 nm CMOS technology," *in IEEE MTT-S Int. Microw. Symp. Dig.*, 2012.

[14] C.-Y. Kuo, A. Y.-K. Chen, C.-M. Lee, and C.-H. Luo, "Miniature 60 GHz slow-wave CPW branch-line coupler using 90 nm digital CMOS process," *Electron. Lett.*, vol. 47, no. 16, pp. 924–925, Aug. 2011.

[15] Y.-C. Tseng and T.-G. Ma, "On-chip X-band branch-line coupler using glass integrated passive device technology," *Electron. Lett.*, vol. 48, no. 25, pp. 1605–1606, Dec. 2012

[16] J.-D. Jin and S. S.-H. Hsu, "A 0.18-µm CMOS balanced amplifier for 24-GHz applications," *IEEE J. Solid-State Circuits*, vol. 43, no. 2, pp.440–445, Feb. 2008.

[17] T.-N. Kuo, Y.-S. Lin, C.-H. Wang, and C. H. Chen, "A compact LTCC branch-line coupler using modified-T equivalent-circuit model for transmission line," *IEEE Microw. Wireless Compon. Lett.*, vol. 16, no. 2, pp. 90–92, Feb. 2006.

[18] Y.-S. Lin and J.-H. Lee, "Miniature Butler matrix design using glass-based thin-film integrated passive device technology for 2.5-GHz applications," *IEEE Trans. Microw. Theory Techn.*, vol. 61, no. 7, pp. 2594–2602, Jul. 2013.

[19] D. Titz, F. Ferrero, R. Pilard, C. Laporte, S. Jan, H. Ezzeddine, F. Gianesello, D. Gloria, G. Jacquemod, and C. Luxey, "New wideband miniature branchline coupler on IPD technology for beamforming applications", *IEEE Trans. Compon. Packag. Manuf. Technol.*, vol. 4, no. 5, pp. 911–921, May 2014.

[20] V. Napijalo, "Coupled line 180° hybrids with Lange couplers," *IEEE Trans. Microw. Theory Techn.*, vol. 60, no. 12, pp. 3674–3682, Dec. 2012.

[21] I. Haroun, Y.-C. Hsu, and D.-C. Chang, "60 GHz rat-race coupler using LG-CPW transmission lines in IPD technology," in *Proc. Int. Topical Meeting on Microw. Photonics & Asia Pacific Microw. Photonics Conf.*, 2011, pp. 284–287.

[22] C.-H. Tseng, "Compact LTCC rat-race couplers using multilayered phase-delay and phase-advance T-equivalent sections," *IEEE Trans. Adv. Packag.*, vol. 33, no. 2, pp. 543–551, May. 2010.

[23] X. Q. Lin, R. P. Liu, X. M. Yang, J. X. Chen, X. X. Yin, Q. Cheng, and T. J. Cui, "Arbitrarily dual-band components using simplified structures of conventional CRLH TLs," *IEEE Trans. Microw. Theory Techn.*, vol. 54, no. 7, pp. 2902–2908, Jul. 2006.

[24] J. Wang, J. Ni, Y.-X. Guo, and D. Fang, "Miniaturized microstrip Wilkinson power divider with harmonic suppression," *IEEE Microw. Wireless Compon. Lett.*, vol. 19, no. 7, pp. 440–442, Jul. 2009.

[25] F.-R. Yang, Y. Qian, R. Coccioli, and T. Itoh, "A novel low-loss slow-wave microstrip structure," *IEEE Microw. Guided Wave Lett.*, vol. 8, no. 11, pp. 372–374, Nov. 1998.

[26] J. Naqui, A. Fernández-Prieto, M. Durán-Sindreu, F. Mesa, J. Martel, F. Medina, and F. Martin, "Common-mode suppression in microstrip differential lines by means of complementary split ring resonators: theory and applications," *IEEE Trans. Microw. Theory Tech.*, vol. 60, no. 10, pp. 3023–3034, Oct. 2012.

[27] A. Fernandez-Prieto, J. Martel, F. Medina, F. Mesa, S. Qian, J.-S. Hong, J. Naqui, and F. Martín, "Dual-band differential filter using broad-band common-mode rejection artificial transmission line," *Progress in Electromagnetics Research*, vol. 139, pp. 779–797, 2013.

[28] C.-H. Tsai and T.-L. Wu, "A broadband and miniaturized common-mode filter for gigahertz differential signals based on negative-permittivity metamaterials," *IEEE Trans. Microw. Theory Tech.*, vol. 58, no. 1, pp. 195–202, Jan. 2010.

[29] T.-W. Weng, C.-H. Tsai, C.-H. Chen, D.-H. Han, and T.-L. Wu, "Synthesis model and design of a common-mode bandstop filter (CM-BSF) with an all-pass characteristic for high-speed differential signals," *IEEE Trans. Microw. Theory Techn.*, vol. 62, no. 8, pp. 1647–1665, Aug. 2014.

[30] C.-Y. Hsiao, C.-H. Tsai, C.-N. Chiu, and T.-L. Wu, "Radiation suppression for cable-attached packages utilizing a compact embedded common-mode filter," *IEEE Trans. Compon. Packag. Manufac. Techn.*, vol. 2, no. 10, pp.1696–1703, Oct. 2012.

[31] B.-F. Su and T.-G. Ma, "Miniaturized common-mode filter using coupled synthesized lines and mushroom resonators for high-speed differential signals," *IEEE Microw. Wireless Compon. Lett.*, vol. 25, no.2, pp. 112–114, Feb. 2015.

[32] L. C. Hsu, Y. L. Wu, J. Y. Zou, H. N. Chu, and T. G. Ma, "Periodic synthesized transmission lines with two-dimensional routing capability and its applications to power divider and couplers using integrated passive device process," *IEEE Trans. Microw. Theory Techn.*, vol. 64, no. 2, pp. 493–501, Feb. 2016.

Index

ABCD matrix, 7, 11, 34, 63, 130, 156, 170, 192
Alcoholic solution, 165
Alternative architecture, 103
  dual-mode retrodirective array, 103
Amplifiers, 102, 115, 119, 122
  bi-directional, 115
  gain, 102, 119, 122
Amplitude imbalance, 52, 122, 128
Appended L-section, 159
Arbitrary outline profile, 194
Array factor, 32, 111
Artificial transmission line, 6
Asymmetric feeding line, 39
Asymmetric network, 28
Attenuation constant, 7, 29, 57, 174
Auxiliary components, 102, 111

Backward-wave directional coupler, 42, 137
Balanced condition, 23
Balanced CRLH, 23
Balanced output, 50
Balanced signal, 50
Balun filter, 41
Bandgap, 20, 23, 169, 186
  common-mode filter, 186
  PBG, 20, 169
Bandpass filter, 26, 38, 41, 127, 142
Bandstop filter, 142, 149
Beam-forming network, 95
Beam-steering, 95, 122
Beam-switching, 21
Beam-switching array, 32, 95, 106, 115, 120, 123
Beam-switching/retrodirective array, 106
Bend, 172, 192
Benzocyclobutene (BCB), 127
Bistatic configuration, 102
Bistatic radiation pattern, 102, 105, 112, 114
  retransmitted pattern, 102, 111, 122
Blass matrix, 32, 95
Bloch impedance, 8, 17, 171, 185, 190
Bloch theorem, 7
Bloch wave analysis, 16, 20, 21, 170
Bond wire, 36, 131
Branch-line coupler, 11, 17, 30, 169
  dual-band, 23
  on the LTCC substrate, 58
  lumped, 13
  shunt-stub loaded, 15
  using periodic synthesized lines, 174, 192
  V-band, 128
Bridged T-coil, 16, 59, 139, 147, 168
Broadside-coupled, 45, 48, 49, 169
Broadside directional coupler, 178
Butler matrix, 26, 32, 95, 106, 111, 118, 139

Capacitive patch, 185, 189
Carrier substrate, 127
  glass, 127
  high-resistivity(R) silicon, 127

Synthesized Transmission Lines: Design, Circuit Implementation, and Phased Array Applications, First Edition.
Tzyh-Ghuang Ma, Chao-Wei Wang, Chi-Hui Lai and Ying-Cheng Tseng.

Category, 65, 75
category-1 design, 66, 73, 75, 91
category-2 design, 66, 73, 75, 79, 91
category-3 design, 66, 70, 75, 83, 91
category-4 design, 66, 71, 75, 87
Ceramic filter, 142
Ceramic layers, 55
Chamfered structure, 192
Characteristic impedance, 4, 8
even/odd-mode, 47, 180
non-dispersive, 12, 20
synthesis equations, 10, 11, 22, 28, 34, 56, 63, 67, 71, 73, 77, 81, 130, 160, 163, 178, 192
Chebyshev
prototype, 144
response, 147
Closely-spaced two-element antenna array, 154
CMOS process, 20, 53, 168
Common excitation, 47
Common input, 90, 117, 153
Common-mode filter, 147, 184
rejection filter, 148, 184
Common-mode noise, 148, 184
Complementary-conducting strip (CCS), 19, 127, 168, 192
Complementary design, 71, 111
beam-switching/retrodirective array, 111
Complex dielectric constant, 151, 165
Complex propagation constant, 7, 29, 186
Composite right/left handed (CRLH), 12
Conformal, 194
Connecting line, 21, 34, 45, 51, 68, 91, 98, 108, 118, 170, 188
Connecting segment, 118
Coplanar waveguide (CPW), 26, 34, 44, 128, 142, 151
lower ground coplanar waveguide (LG-CPW), 128
Corner cell, 192
Coupled line, coupled line section, 39, 42, 49, 137
synthesized coupled line, 42, 49
Coupled-line coupler, 137
Coupled line theory, 43
Coupled port, 12, 44, 54, 139, 181
Coupled slotline mode, 36
Coupling coefficient, 44, 47, 50, 53, 139, 147, 180
Crossed conducting strip, 169
Cross-sectional profile, 38, 127, 181
broadside coupler, 181
integrated passive device (IPD), 127
Cubic equation, 156
Current wave, 4, 5, 7
Cutoff frequency, 12, 15, 18, 36, 150, 174

DC block, 42
Decoupling network, 127, 151, 154
Degenerate modes, 39
Degree of freedom, 68, 85, 109, 118
three degrees of freedom, 160, 186
two degrees of freedom, 56, 71
Differential circuit, 50
Differential excitation, 47
Differential signal, 147, 184
Diplexer, 63, 68, 87, 117, 124, 151
Directional coupler, 26, 42, 44, 53, 128, 137, 168, 178
backward-wave, 42
broadside-coupled, 49, 178
edge-coupled, 45
forward-wave, 168
lumped, 53
on-chip design (IPD), 139
Direction of arrival, 95, 103, 107, 112, 122
Direct-thru, 98, 109, 117
Discontinuities, 30, 33
Dispersion, 185
diagram, 23, 187
Dispersive, 12, 20, 26, 36, 75, 80, 174
Distributive, 70
Drain mixer, 103
Dual-band hybrid, 109
Dual-band operation, 96
Dual-mode, 38
balun bandpass filter, 41
bandpass filter, 38
coupler, 97, 112, 124
crossover, 98, 108, 112
delay line, 112, 124
feeding network, 102, 106, 109, 113
integrated beam-switching/retrodirective array, 106
integrated phased array, integrated array, 69, 114, 122
left-handed synthesized line, 111
line, 65, 69, 71, 73, 75, 124, 154
operation, 73, 109, 115, 161
resonator, 38
retrodirective array, 73, 96, 103
retrodirective/beam-switching array, 111, 124
synthesized line, 66, 70, 72, 87, 91, 109, 152, 154
Dual-operational mode, dual-mode, 63, 65, 72
synthesized transmission line, 65

Edge-coupled coupler, 45
Effective dielectric constant, 47, 181
Eight-port network, 98
S-parameters, 101
Electrical length, 5, 9, 26, 91, 117, 128, 144, 168
synthesis equations, 56, 63, 73, 77, 81, 86, 130, 150, 156, 163, 171, 178
Electric field polarization, 102
Electrodes, 6, 34, 37, 127
Element spacing, 32

Enveloped correlation coefficient (ECC), 124
Equal-ripple Chebyshev response, 147
Equivalent circuit model, 63, 66, 76, 79, 84, 153, 186
  equivalent lumped model, 27, 55, 72, 137, 178
  equivalent model, 34, 66, 67, 70, 145, 146, 152, 185
Even mode, 46, 50, 144, 149, 179, 185
  characteristic impedance, 44, 47, 49, 178, 180, 186
  complex propagation constant, 188
  dispersion diagram, 187
  effective dielectric constant, 48, 181
  half circuit, 40, 43, 47, 149, 179, 185
  inductance and capacitance, 46
  input admittance, 133
  input impedance, 40
  phase constant, 50, 187
  phase velocity, 43
  Z-matrix, 187
Even-odd mode analysis, 40, 43, 46, 50, 143, 148, 178
Even-/odd-mode
  characteristic impedance, 47, 50, 180
  half circuits, 149
Experimental setup for bistatic radiation patterns, 102
Eye diagram, 150, 188

First-order equivalent model, 34
Five-element network, 64, 65
  Five-element ladder-type network, 56, 63, 67, 72, 76, 80, 85, 135, 154, 161
Folding method, 98
Forward-wave directional coupler, 168
Four-element array, 122
  Butler matrix, 107
  dual-mode retrodirective array, 98, 103
  heterogeneous integrated array, 106
  reflection-type RDA, 107
  retrodirective circuit, unfolded, 99
  Van Atta array, 118
Four-port
  even-mode Z-matrix, 187
  network, 47, 97, 154, 178
  S-parameters, 47
Four-port network, 47, 97, 154, 157, 178
Free space wavelength, 59
Free space wavenumber, 5, 32
Frequency selective element, 157
Frequency selectivity, 39

Guided wavelength, 17, 28, 40, 50, 95, 145
  losses per guided wavelength, 30, 36
Guided wavenumber, 4, 5, 28

Half-circuit model, 63, 64, 149, 179, 186
Harmonic, 27, 40, 168, 174
  harmonic suppression, 168

Heterogeneous integrated phased array, 34, 62, 95
  simplified dual-mode, 122
  tri-mode, 115
High-ε_r, 127
Highpass, 24
High-speed differential system, 148
Host line, 7, 168, 190

IF (intermediate) signals, 103, 122
Illuminating antenna, 102, 103, 122
Illuminating direction, 103
Impedance transformer, 80, 157
In-band, 40, 62, 89, 145
Incident angle of illuminating wave, 111, 112, 122
Inductor
  meander-line, 13, 27, 34, 70, 130, 184
  spiral, 13, 127, 138, 145
  vertical, 26, 55, 59
Infinite input impedance, 67, 73, 83, 88, 157, 163
Inner product, 38
Input reflection coefficient, 68, 71, 73, 79, 83, 87, 161
Integrated passive device (IPD), 10, 126, 127, 169, 189
Integrated phased array, 69, 114, 115, 122
Interdigital capacitor, 6, 27, 45, 51, 72, 150
Interrogation signal, 95, 98, 102
Interrogator, 95, 107, 115
Isolation, 21, 38, 48, 59, 89, 91, 101, 109, 117, 131, 139, 157, 161, 174, 178, 181
  isolation port, 54, 123
Iterative process, 48

J-inverter, 143, 144

Kirchhoff's laws, 3

Ladder-type model, 63, 64, 77, 81
Ladder-type network, 56, 63, 65, 67, 70, 73, 76, 84, 87, 135, 154, 161
  five-element, 56, 63, 67, 72, 76, 80, 85, 135, 154, 161
  nine-element, 64, 87
  three-element, 63, 70, 77, 81, 147
LC tank, 27, 63, 64, 67, 76, 131, 134, 153, 163, 186
  LC resonator, 21, 72, 109, 134, 159, 185
Leaky-wave antennas, 21
Left-handed, 1, 13, 21, 59, 62, 66, 70, 73, 111, 151, 168
  passband, 23
  phase-advance, 59
  structure, 21, 87
  synthesized CPW, 151
  synthesized line, 1, 22, 89
Line equivalence, 4, 7, 12, 13
Line resolution, 126, 173
Liquid detector, 151, 161, 165
Local oscillator, 95, 103, 115, 118
LO signal, 122

Lossless
assumption, 22
distributed uniform transmission line, 3, 4
periodic LC network, 17
property, 171
series LC tank, 134
transmission line, 4, 5, 6, 8, 12
Losses per guided wavelength, 30, 36
Lossy transmission line, 10, 11
Lower ground CPW (LG-CPW), 128
Lowpass, 6, 12, 41
Low temperature co-fired ceramic (LTCC), 55, 188
filter, 142
L-section, 28, 34, 47, 67, 84, 159
L-section model, network, 29, 159
Lumped balun, 53
Lumped coupler, 13, 54
Lumped-element portion, 77, 80, 83, 156
Lumped network, 1, 6, 11, 27, 62, 76, 80, 168, 192
non-periodic, 62, 192
quasi-lumped network, 1, 6, 16, 27, 62, 127, 159

Marchand balun, 26, 42, 49, 53
Matching unit, 89, 153, 157
Material medium, media, 1, 6, 161
Maxwell's equations, 1, 2
Meandered vertically, 55
Meander high impedance line, 6, 28
Meander-line inductor, 27, 34, 70, 130, 184
MEMS (microelectromechanical system), 111
Metal-insulator-metal (MIM), 13
capacitor, 55, 127, 130, 138, 143, 145, 172, 189
Metal-oxide-metal (MOM), 13
Metamaterial, 1, 13, 21
Microstrip line, 21, 23, 27, 76, 128, 172, 192
high-impedance, 41
Middle stage, 64, 68, 138
Mode conversion loss, 36
Mode splitting, 39, 40
Modified-T circuit, 59, 148
Multilayer, 53, 55, 131, 169, 181, 189
Multi-mode
operation, 24, 127
synthesized CPW, 151
synthesized line, 62, 63, 64, 119
Multi-operational mode synthesized transmission line, 2, 62, 92, 96
Multiple stubs, 15
Multiplexer, 62, 127, 151
Mushroom resonator, 184, 188
Mutual capacitance, 49, 53, 139, 144, 147, 150, 168, 181, 185
mutual capacitor, 48, 138, 139
Mutual inductance, 45, 139, 147, 150, 181, 185
mutual inductor, 138

Narrowband, 10, 15, 129, 168
N-element linear array, 32
Nine-element ladder-type network, 64, 87
Nine-pole network, 85
90° line, 30, 36, 83, 128, 134, 169, 172, 190
90° synthesized line, 55, 57
Node exchange, 28, 56, 64, 77, 81, 85, 130
Nolen matrix, 32, 95
Non-integer ratio, 21
Non-periodic structure, 168
Non-periodic synthesized line, 15, 27, 30, 41, 55, 59, 130, 139, 146, 168
Non-resonating LC tank, 159
Notch structure, 192

Odd mode, 43, 46, 143, 148, 179, 181, 185
Bloch impedance, 187
characteristic impedance, 44, 47, 50, 178, 180, 186
dispersion diagram, 187
effective dielectric constant, 48, 181
half circuit, 40, 43, 47, 149, 179
inductance and capacitance, 46
input admittance, 133
input impedance, 40
phase constant, 50, 187
phase velocity, 43
On-chip implementation, realization, 16, 127, 189
One-dimensional periodic structure, synthesized line, 9, 16, 18, 169
One-dimensional wave equation, 3
Open circuit condition, 65, 73, 79, 82, 86, 153, 157
Open-circuited, 65, 69, 73, 77, 81, 85, 87, 91, 98, 108, 131, 152, 164, 185
Open-ended stub, 63, 70, 99, 128
Open stub, 15, 18, 39, 70, 91, 97, 99, 127, 170, 173, 185, 190
loaded equivalence, 14
Operating scenario, 75
Orthogonal modes, 38, 154
Out of band, 62, 89
Output phase difference, 58, 139, 177, 183, 193
Over-coupled, 53

Parallel LC resonator, 21, 72, 185
Parallel LC tank, 68, 86, 131, 164
Parallel-plate capacitor, 27, 34, 41, 57, 72, 76, 80, 84
Parallel-plate waveguide, 2, 4
Parallel resonator, 63, 65, 72, 147
Parasitic loss, 13, 26, 45, 56, 102, 165
Passband, 21, 39, 63, 67, 70, 75, 80, 91, 145, 147, 153, 157, 186
ladder-type approximation, 63
Pattern generator, 150
Periodic interdigital stub, 168

Periodicity, 1, 6, 7, 8, 10, 20, 22, 26, 36, 53, 75, 80, 170, 171, 172, 178, 181, 185, 189
Periodic structure, 9, 16, 21, 168
Periodic synthesized line, 16, 18, 19, 169, 172, 174
 coupled, 178, 180
Permeability, 3, 5, 21
Permittivity, 3, 5, 21
Perturbation, 38, 41
Per-unit-cell
 capacitance, 36
 even-and odd-mode inductances and capacitances, 46
 inductance, 36
 phase delay, 190
Per-unit-length
 capacitance, 3, 6
 inductance, 3, 6
Phase adjustment, 32, 76, 80, 95, 97, 103, 112, 172
Phase-advance section, 59
Phase compensation, 51, 108, 112, 118
Phase-conjugating array (PCA), 95, 103, 115, 118, 122
Phase constant, 4, 5, 8, 34, 171
Phased array, 1, 32, 69, 83, 95, 106, 108, 161
 dual-mode, 96, 103, 106, 111, 124
 tri-mode, 115
Phase delay, 26, 29, 59, 79, 83, 87, 98, 118, 124, 168, 172, 190
 section, 59
Phase difference, 12, 15, 18, 31, 32, 38, 42, 49, 52, 101, 111, 118, 122, 131, 157, 174, 176, 178, 181
 output phase difference, 58, 139, 177, 183, 193
Phase difference reversal condition, 95, 101, 107, 118
Phase imbalance, 34, 42, 48, 51, 53, 122, 136, 158, 174
Phase inverter, 129, 131
Phase linearity, 12, 26, 57
Phase velocity, 3, 4
Photonic bandgap (PBG), 20, 169
PIN diode, 111
π-equivalence, 127
π-model, 9, 14, 16, 146
π-network, 11, 13, 28, 53, 70, 76, 146, 156, 168
π-resonated varactor, 55
Planar crossover, 90
Plane of symmetry, 50, 84, 144, 153, 185
 symmetric with respect to the central plane, 76, 79, 133, 143, 148, 178
+-shaped strip, 189, 190
Positioning, 106, 111, 115
Post-integration tuning, 29, 36, 45, 86
Printed circuit board (PCB), 13, 26, 32, 48, 126, 165, 169, 181, 184, 189
 double-sided, 44
Progressive phase difference, 32, 118
Propagation constant, 7, 8, 11, 16, 17, 34, 47, 186
Propagation delay, 150
Pseudorandom binary sequence (PRBS) differential signal, 150, 188

Quadratic function, 144
Quadrature hybrid, 11, 16, 30, 32, 54, 90, 95, 97, 107, 113, 117, 128, 139, 154
Quality factor, 7, 20, 56, 68, 73, 85
 external quality factor, 147
 unloaded quality factor, 26, 36, 40, 45
Quarter-wavelength, 89
 coupled lines, 50
 dual-band, 133, 135
 line, 14, 129, 133, 172, 189, 192
 periodic synthesized line, 169, 172, 174
 synthesized lines, 12, 18, 27, 176, 189
 transformer, 83, 98
Quasi-lumped
 component, element, 28, 36, 49, 76, 86, 92, 127, 145
 network, 1, 6, 16, 27, 62, 127, 159, 168
Quasi-TEM, 5

Radar and image systems, 1
Radiation loss, 36
Radiation pattern, 32, 111, 114, 121, 124
 bistatic, 102, 105, 112, 114
 retransmitted, 102, 111, 122
Rat-race coupler
 dual-band, 132
 dual-mode, 154
 LTCC, 59
 miniaturized, 30, 37
 wideband, 129
Reactive loading, 102, 111, 133
Realizable frequency ratio, 94
Receiving array, 107, 120, 122
Reciprocal, reciprocal network, 8, 11, 171
Reentrant signal, 107, 122
Reflection-type phase shifter, 26, 54, 98
Reflection-type retrodirective array, 95, 97
Reflective load, 54, 55, 95, 107
Relative dielectric constant, complex, 161, 164, 165
Requisite condition, 68, 76, 85, 156
Retransmission, 98
Retransmitted (bistatic) pattern, 102, 105, 111, 112, 114, 121, 122, 124
Retrodirection, 118, 122, 124
Retrodirective array (RDA), 73, 95, 96, 105, 123
 dual-mode, 73, 96, 103
 phase conjugating, 95, 103, 115, 118, 122
 reflection-type, 95, 97
 Van Atta, 95, 98, 111, 115, 118, 122
Retrodirective/beam-switching array, 111

Return path, 147
RF signal, 122
Right-handed, 1, 21, 62, 139, 168
  connecting line, 21
  five-element ladder-type network, 72
  passband, 23
  phase-delay, 59
  properties, 71
  synthesized CPW, 151
  synthesized line, 66, 89, 117
Ring resonator, 26, 39, 41

Sawtoothed coupling structure, 39, 40
Scanning range, 124
Self capacitance, 49, 180
Self inductance, 53
Sensitivity coefficient, 165
Sensor network, 115
Series element, 63, 102
Series impedance, 17, 73, 159, 160
Series inductance, 6, 28, 34, 45, 53, 130, 156, 170, 189
Series LC resonator, 21, 109, 134
Series LC tank, 64, 67, 76, 80, 84, 134, 153, 161
Series resonator, 66, 68, 74, 75, 83, 134, 149, 154
Short circuit condition, 71
Short-circuited, 50, 64, 65, 70, 91, 164
Shunt admittance, 159, 161
Shunt capacitance, 6, 20, 34, 41, 45, 50, 53, 55, 70, 128, 130, 156, 170, 186, 190
  fictitious, 143
  of inductance (parasitic), 27, 57, 159
Shunt element, 7, 63, 64, 102
Shunt-stub, 17
  loaded, 14, 16, 17
Sidelobe, 34, 103, 111
Signal attenuation, 30, 42
Signal flow, 51, 89, 98, 99, 103, 112, 117
Signal-to-interference-plus-noise ratio, 32
Simplified dual-mode integrated array, 122
Simultaneous equations, 68, 71, 164, 171, 186, 189
Single-ended input, 41, 51
Size reduction ratio, 13, 28, 31, 37, 41, 58, 131, 135, 168, 174, 176, 192
Slope parameter, 144
Slow wave
  characteristics, 21
  loading, 41, 53
  synthesized CPW, 36, 39, 130, 143
  synthesized lines, 6, 159
Slow wave factor (SWF), 5, 13, 20, 28, 29, 30, 36, 45, 57, 131, 174
Slow-wave structure, 35, 48
Smart living space, 115
Smith chart, 69, 71, 73, 83, 87

S-parameters, extraction of line parameters, 10
SPDT switch, 120
Spiral coils, 53
Spiral high-impedance line, 6, 28
Spiral inductor, 127, 138, 145
SP4T switch, 120
Stepped-impedance resonator (SIR), 143
Stopband, 39, 63, 65, 67, 70, 143, 186, 188
  attenuation, 31, 71
  extended, 143, 145
  response, 62
  Upper stopband, 145
Stripline form, 48
Sum and difference patterns, 123
Surface acoustic wave (SAW) filter, 142
Surface-mounted device (SMD), 13, 67, 71
Switching circuit, 107
Symmetric network, 9, 28
Synthesized CPW (SCPW), 34, 37, 39, 128, 142, 158
  coupled, 42, 46, 50, 137
  double-sided, 38
  dual-mode, 155, 163, 165
  left-handed (RH), 151
  multi-mode, 151
  non-periodic, 34, 128
  right-handed (RH), 151
  slow wave, 36, 39, 130, 143
Synthesized microstrip line, 27, 29, 32
Synthesized stepped impedance resonator, 143
Synthesized transmission line, synthesized line, 6
  non-periodic, 15, 27, 30, 41, 55, 59, 130, 139, 146, 168
  periodic, 16, 18, 19, 169, 172, 174
System equations, 35, 44, 79, 82, 86
System-in-package (SiP), 55, 126
System-on-chip (SoC), 126

Telegrapher's equations, 3
T-equivalent model, 28
Thin-film resistor, 127
Three-channel selector, 116, 117
Three-dimensional routing, 55
Three-element network, 9, 10, 55, 63
Three-quarter wavelength line, 129, 131
Three-quarter wavelength synthesized line, 176
Thru port, 12, 54
Tight coupler, 44, 127, 128
T-model, 9, 14, 15, 16, 17, 24, 47, 170
T-network, 17, 34, 45, 130, 139, 168, 185
Transformer-coupled bandpass filter, 146
Transmission channel, 88, 91, 117, 122, 157, 159
Transmission line model, 3, 16, 40
Transmission-line portion, 80, 84, 156, 157
Transmission path, 91, 101, 117, 122, 151, 157, 192

Transmission zero, 27, 30, 40, 64, 67, 70, 77, 80, 153, 184
(frequency), equation of, 29, 57, 65, 68, 73, 78, 82, 86, 150, 159, 163
parasitic transmission zero, 57
second-order transmission zero, 39
single-order transmission zero, 39
split, 71
three transmission zeros, 145
two transmission zeros, 83, 147, 149
Transmitting array, 106, 107, 122
Transponder, 106, 115
Transverse electromagnetic (TEM), 2, 5, 12, 43, 179
Tri-mode
beam-switching/Van Atta/ phase conjugating array, 115
coupler, 117
feeding network, 119, 122
heterogeneous integrated phased array, 96, 115
operation, 83, 168
synthesized line, tri-mode line, 62, 74, 75, 79, 83
Tri-operational mode synthesized transmission line,
Triplexer, 63, 71, 75, 80, 83, 89, 117, 157
TRL (thru-reflect-line) calibration, 172
Tuning process, 92, 174, 187
post-integration, 29, 36, 45, 86
Two-dimensional routing, 9, 19, 168, 169, 172, 188, 192
Two-element array, 123
Two-element retrodirective/beam-switching array, 124
Two-fold symmetry, 43, 117
Two-port
half circuits, 43
network, 54, 149
S-parameters, 10, 11, 29, 43, 47
Y-parameters, 140
Z-parameters, 16, 140

Ultra-thick metal, 127
Unbalanced CRLH, 23
Unbalanced signal, 50
Uncoupled CPW, 45
Uncoupled transmission lines, 108, 154, 157
Uniform amplitude excitation, 32
Uniform transmission line, uniform line, 1, 6, 7, 9, 14, 34, 62, 128, 171
Uniplanar, 45, 49
Unit cell of, 7, 22, 48, 169, 171
bridged-T coil, 148
CCS, 19
CRLH, 22
one-dimensional periodic structure, 16
periodic synthesized line, 169, 172, 180, 184, 189
synthesized CPW, 36
tri-mode synthesized line, 75

Van Atta array, 95, 98, 111, 115, 118, 122
Variable attenuator, 55
Vertical inductor, 26, 55, 59
Via hole, 34, 55
Virtual short circuit, 64, 67, 68, 86, 91, 145
Voltage wave, 4, 5, 7

Water volume fraction of solution, 165
Wave equation, 2, 5
Wave-guiding structure, 5, 24, 34
coupled wave-guiding structure, 42
Wavelength
free space, 59
guided, 17, 28, 30, 36, 40, 50, 95, 145
Wheeler's cavity model, 38
Wilkinson power divider, 169, 174, 192
Wireless charging, 115
Wireless power transmission, 96, 106, 115
Wireless sensor network, 115

x-shaped strip, 189, 190

Y matrix, 10, 133
Y-parameters, 140

Z matrix, 10, 187
Z-parameters, 16, 141